Praise for *Understanding Li*

"The smartest guy I know has written the most illuminating book of the year. Now if I can just summon the discipline to follow his advice!"

MITCH DANIELS, president, Purdue University, and former governor of Indiana

"Don Brown has taken an incredibly diverse range of topics—origins of the universe and life, evolution, biology, chemistry, disease progression, and how to live better lives—and connected them together in a purposeful way. He has a true gift of making the complex understandable through stories, pictures, and analogies. It was amazing to see how he links the activities of 4.5 billion years of evolution together in a way that relates to each of us as individuals and our own health. I know this will change the way I think and live."

DENNIS MURPHY, president and CEO, Indiana University Health

"Dr. Brown masterfully makes an impossibly complex topic accessible. Beautifully written!"

MARC COHEN, cofounder and executive chairman, C4 Therapeutics

"I found *Understanding Life* an interesting and fun read, as it presented a foundational explanation of the underpinnings of the science that allowed life to evolve in a manner that anyone can understand. One does not have to have a background in science or medicine to 'get it.' I recommend *Understanding Life* to anyone who wants to understand how and why we are here."

DICK RECK, president, Business Strategy Advisors LLC

"Don Brown strikes again, this time addressing the beautiful complexities of life itself. As someone who's devoted himself to both improving the present and challenging the status quo, in less than three hundred pages, Don provides an answer key to improving the quality of our personal and professional lives. Fortunately, this one book does the research for you and peels back multiple layers of history, conveniences, and personal decisions that lead to the condition of our health and wealth no matter what your age, stage, or place you call home. Now comes the hard part—applying the research to the daily decisions we make. This page-turner, *Understanding Life*, has the power to change your *life* for the better. It's been a gift to obtain, and I intend to pass it on."

ERIC HOLCOMB, governor, State of Indiana

"A timely, must-read book for anybody interested in being their healthiest self. Dr. Brown artfully simplifies complex scientific and medical information, empowering readers to take real control over their health and well-being."

ED SIMCOX, vice president, Mayo Clinic, and former chief technology officer, U.S. Department of Health and Human Services

Understanding Life

Understanding

DON BROWN, MD

Life

Tap Into An Ancient Cellular
Survival Program to Optimize
Health and Longevity

This book is not intended as a substitute for the medical advice
of physicians. The reader should regularly consult a physician in
matters relating to his/her health and particularly with respect to
any symptoms that may require diagnosis or medical attention.

Cataloguing in publication information is
available from Library and Archives Canada.
ISBN 978-1-77458-156-8 (paperback)
ISBN 978-1-77458-157-5 (ebook)

Page Two
pagetwo.com

Edited by Scott Steedman
Copyedited by Rachel Ironstone
Cover design by Taysia Louie
Interior design by Fiona Lee
Illustrations by Tori Rogers

lifeomic.com/team

All author proceeds go to the Riley Hospital
for Children at Indiana University Health.

To my children and grandchildren,
who continually challenge me
to see the world through fresh eyes.

Contents

. . . .

Introduction

T HIS BOOK is an almost ridiculously ambitious undertaking.
It attempts to span an enormous conceptual range, from the
development of cellular life on our planet to health, aging, and
disease. It covers topics ranging from biochemistry to human
physiology and moves from the theoretical to the practical. At the
outset, when I described my plans, many people openly scoffed
at the notion that a casual reader with little or no scientific background could understand such things—or even want to. However,
as much as anything, this book reflects my faith in the average
person. I don't believe you need a doctorate to grasp the principles underlying biology. Further, I think that all of us have a deep
fascination with how living things work and what it means for our
own health and well-being. I've done my best to simplify the more
complex topics and to avoid scientific jargon as much as possible.

I'm afraid that there's no way to explain the development of
life without invoking the potentially controversial "e-word"—
evolution. Let me take a few moments to try to convince you that
it's OK for even the most ardent creationist to believe in evolution,
or at least the evolutionary process as we observe it today, and
recognize its predictive value.

Have you ever fired a shotgun? The shells that are loaded into
such weapons are filled with hundreds of tiny metal balls, appro-
priately called shot. If you fire a shotgun at a target across a field,

you'll find it peppered with little holes that reflect how fast each piece of shot was going and its exact trajectory when it hit the target. The farther you are from the target when you shoot, the more spread out the shot will be, reflecting the dispersion that occurs over time, as demonstrated in figure 1.

Fig. 1: Shooter/Target

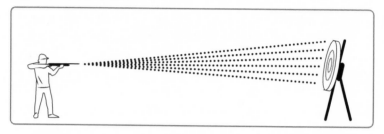

If someone gave you a target they had fired at, can you see how it would be possible to work backward and figure out where they were standing when they fired the gun? Such analysis would involve fairly simple physics using math that Isaac Newton developed more than three centuries ago. You could probably do a fairly good job just by eyeballing the target and the amount of dispersion without using any math at all.

However, if you didn't see the person fire the gun, could you ever be *sure* where they were standing? What if someone developed a machine that precisely mimicked the position, angle, and speed of every single piece of shot halfway between the hypothetical shooter and the target, as shown in figure 2?

Fig. 2: Shooter/Machine/Target

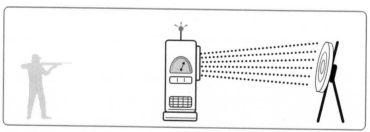

When it leaves the machine, the shot will have the exact overall state (position and velocity) as if it had been fired by a person holding a shotgun at the calculated spot. Newton's calculations would be accurate in either case, but the point of origin would be different.

The situation with evolution and creation is similar. Evolution is analogous to Newtonian mechanics in the shotgun example above, as something that can be measured and verified. In the same way, we see evolution at work when bacteria become resistant to antibiotics. Even cancer is an example of evolution in action. So even though we talk about evolution as a *theory*, it is extremely well established. We also have a *theory* for electromagnetism that similarly works so well that it has led to smartphones and other sophisticated devices. You don't have to accept the theory to use your iPhone to call your mother.

The controversy begins when we run the theory of evolution in reverse and use it to explain the origin of life. Just as with the shotgun example, we can never know for sure. After all, no one was around when the earth was formed or when life began. In the same way that, given a target, we can estimate where the shooter was but can't rule out the possibility that the shot pattern was generated elsewhere that could replicate the same state as a shooter, we can't disprove that the earth and its life arose from some miraculous event just a few thousand years ago. Either way, the theory of evolution is just as valuable. In fact, if there's anything divine and perfect about life, it's the process of evolution. If you look closely at living things, it's clear that they are not the perfectly designed products of some super intelligence, but marvelous Rube Goldberg contraptions that work for a short period of time and then fall to pieces.

So bottom line—it's OK for even the staunchest creationist to believe in the process of evolution and to recognize its predictive value. Scientists may choose to run the models backward and talk in terms of how things appear to have begun, and religious people may prefer to believe that the world was created by a divine being much more recently. However, both can apply evolution practically to help understand how life, bacteria, cells, and organisms operate. A quote from Theodosius Dobzhansky sums the situation up nicely: "Nothing in biology makes sense except in the light of evolution." It really is something we can all agree on despite our different faith backgrounds. In this book, we'll go ahead and run the models of evolution backward. This will help us understand the "apparent" development of life because it will allow us to appreciate how our bodies operate.

This book is organized into three multi-chapter sections. The first examines the molecular basis of life and how it might have emerged on the early earth. The second explores aging and death in complex organisms—like us. The third section looks at some of the ancient programs that our cells inherited from their primitive ancestors and how we can leverage these programs to live longer and healthier lives.

There's no denying that some of the material here may be challenging for people who didn't study chemistry and biology in college. For that reason, I have partnered with LifeOmic to create easy-to-consume mobile courses that distill the essence of each chapter into short comic books that can be accessed using the Lifeology platform. Lifeology has been used to create dozens of illustrated courses focusing on various aspects of science and medicine. If you really want to understand the information presented in this book, we recommend that you take the following approach:

1 Go through the Lifeology course(s) for a chapter.
2 Read the chapter in the book.
3 Go back through the Lifeology course(s) for the chapter.

The Lifeology courses feature beautiful explanatory drawings by professional artists. These illustrated courses are intended to make learning both easy and fun. You can find the courses for this book online at lifeology.io/understanding-life.

PART ONE

The Molecular Basis of Life

1

The Grand Timeline of Life

"Tune your television to any channel it doesn't receive and about 1 percent of the dancing static you see is accounted for by this ancient remnant of the Big Bang. The next time you complain that there is nothing on, remember that you can always watch the birth of the universe."

BILL BRYSON

IF WE run the various models of physics and chemistry in reverse, all lines converge at a strange event that occurred 13.8 billion years ago, starting with the "big bang." At that instant, the entire universe—all of matter and space itself—was compressed into a volume smaller than the head of a pin. At this pressure and temperature, there were no atoms or molecules, let alone stars or planets. All that existed was energy and fleeting subatomic particles. Space itself began to expand. As it cooled, the contents of the universe began to settle into tiny particles such as quarks and leptons. These soon gave rise to protons and electrons, which in turn began to form atoms—initially hydrogen and helium. We'll go quickly through a crash course in chemistry in the next chapter, but now let's try to get the whole big picture straight in our minds.

As the cloud of hydrogen, helium, energy, and particles spread, these substances began to appear throughout the expanding universe. They became unevenly distributed, and some aggregated into giant clouds of gas and dust that we call nebulae. In many places, these gas clouds began to spin and over the course of several million years started to aggregate into balls. In our neck of the woods, in the Milky Way galaxy, a solar system formed about 4.6 billion years ago with a central star (our sun) and eight planets.

Around 99.9 percent of the mass became concentrated in the area of the sun, leading to tremendous temperatures and pressures. These quickly built up to the point that hydrogen atoms began to fuse together. Some of the mass was converted into energy, leading to the chain reaction we call nuclear fusion. And so the sun began to shine. Even after all these billions of years, hydrogen atoms are still fusing and generating the tremendous energy emitted by our local star.

The small quantity of matter not sucked into the sun settled into the eight planets that began to orbit the new star at various distances. Violent solar winds blew the light hydrogen and helium atoms away from the inner planets, leaving them mostly rocky with thin atmospheres. The four outer planets became the gas giants we know as Jupiter, Saturn, Uranus, and Neptune. The third planet happened to form in an ideal location about ninety-three million miles (150 million kilometers) from the sun. This Goldilocks distance was neither too hot nor too cold. The newly formed planet was able to hold on to an atmosphere of mostly nitrogen and carbon dioxide, and to develop and retain an ocean of water.

The First Life Forms: Bacteria and Archaea

We'll talk later about how life might have originated, but there is overwhelming evidence that this occurred about four billion years ago, roughly a half-billion years after the planet cooled. The first forms of life we know of were the single-celled organisms we call bacteria. These creatures were relatively simple and still are today. We can most succinctly describe them as little protein factories surrounded by a fatty (lipid) membrane. The proteins they make are themselves tiny machines that can perform an amazing variety of wonderful tricks. Some proteins facilitate chemical reactions and are known as enzymes. Others help the bacteria move around, while still others act as harbors, transporting material in and out of the cell through the enclosing membrane.

How do the bacterial cells know what proteins to make and how to make them? That's the job of their DNA. You can think of DNA as a recipe book for proteins. There's one recipe for a protein that can pump sodium out of the cell and another for a protein that can help repair the DNA if it gets damaged. Early "protocells" probably started out with only a handful of proteins. The first true bacterial cells had roughly five hundred or so such recipes, producing proteins capable of handling a range of different jobs. Even after four billion years of evolution, modern bacteria are quite similar to these early ones, now making as many as several thousand different proteins.

Bacteria have been spectacularly successful, spreading out all over the earth. They came into being at a time when there was very little free oxygen (O_2) in the air or dissolved in the water. In fact, oxygen was a poison to these new creatures. For millions of years, bacteria continued to evolve, mainly to explore different ways of harvesting energy to survive and multiply. After about a half-billion years of existence, a new, tougher kind of single-celled creature came onto the scene as it evolved from bacteria. If bacteria were Honda Civics, these new creatures were armored tanks. They were able to live in places that bacteria couldn't even visit— hot springs with temperatures above boiling; polar regions that stayed frozen year-round; lakes so salty that you could float without effort; even water so acidic it would melt your boots. These new, badass bacteria were discovered only in the last fifty years and were given the name *archaea* (singular *archaeon*), which derives from the Greek word for ancient. Interestingly, scientists have never found an archaeon that makes humans sick, unlike their bacterial relatives.

Under the microscope, bacteria and archaea look pretty much the same—just a tiny rod or sphere about one micron (millionth of a meter) across. Archaea can be slightly larger, but not enough to worry about; you'd have to put ten to twenty thousand of them end to end to make an inch. Bacteria and archaea are both *prokaryotes*. That is, they lack a nucleus. Their DNA floats in the interior of the cell, the *cytoplasm*, along with everything else.

Fig. 3: Bilayer structure

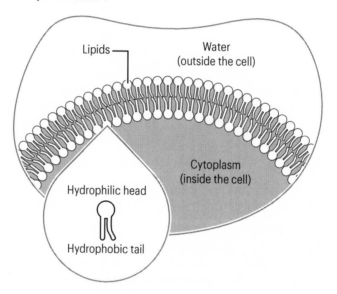

These two types of primitive organisms differ in less apparent ways. For one thing, there are important differences in the ways they copy their DNA when they get ready to divide—archaea use copying machinery much like that of more modern organisms such as plants and animals. The other big difference is in their cell membranes. Bacteria use a bilayer membrane structured like a sandwich. Each layer consists of fatty acids with their charged (polar) heads all grouped together. The long tails of the fatty acids hate water (hydrophobic) and stick together in the middle of the membrane—the inside of the sandwich. The polar heads love water (hydrophilic), so they form the top layer of the sandwich, polar heads sticking out to face the outside world, and the bottom layer of the sandwich, polar heads pointing down to face the watery inside of the cell (cytoplasm).

This structure allows bacterial membranes to be soft and to change shape easily. On the other hand, archaeal membranes can sometimes use lipids that run the full width of the membrane. It's as if the two pieces of bread in the sandwich were held together with toothpicks. It's easy to see how this helps archaea handle more extreme conditions. Instead of being surrounded by a delicate soap bubble like bacteria, archaea are encased in something akin to clear plastic—much tougher and more durable.

Fig. 4: Archaeal membranes

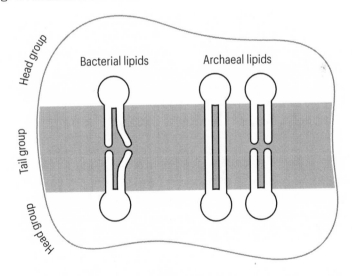

So bacteria arose about four billion years ago, and archaea appeared roughly a half-billion years later. For the next billion years, the picture of life on earth was relatively static. Bacteria spread around the planet into all but the most inhospitable areas. The archaea occupied the places that bacteria feared to tread. Together, they dominated the planet and still do today.

An Amazing New Trick

Around 2.5 billion years ago, one type of bacteria learned how to harness sunlight. Actually, this discovery might be even older, but evidence abounds that at least 2.5 billion years ago some bacteria figured out how to use sunlight to crack hydrogen atoms and their electrons out of water and use them to generate energy. Now, stealing hydrogen atoms and electrons from water is no mean feat because it puts you in conflict with a chemical bully—oxygen. We'll talk about this a lot more in the next chapter, but know for now that oxygen simply loves electrons. Once it gets them, it hates to give them up. So it took a long time—a billion years in fact—for some bacteria to stumble onto a way to reliably accomplish this theft. This ability to use two plentiful resources—sunlight and water—to produce energy was a game-changing innovation. Suddenly, these photosynthetic bacteria began to proliferate, especially in the top layers of the ocean, where sunlight was plentiful. One fascinating wrinkle to this development was the fact that molecular oxygen (O_2) began to accumulate first in the ocean and later in the atmosphere.

Of course, we humans tend to think of oxygen as the elixir of life. After all, all the animals we're familiar with rely on oxygen to live. However, when photosynthetic bacteria burst onto the scene roughly 2.5 billion years ago, the oxygen they spewed out was like the most toxic environmental spill ever recorded. Although later life learned how to actually use oxygen to produce energy and grow, the bacteria and archaea present 2.5 billion years ago had no idea how to deal with the stuff. For them, oxygen was a lethal poison that interfered with some of their most critical chemical reactions. They and their ancestors had spent hundreds of millions of years in an environment with almost no free oxygen. As the photosynthetic bacteria started to proliferate, countless other bacteria and archaea began to die off—or retreat to the deeper ocean, still devoid of the deadly poison.

At first, the oxygen the photosynthetic bacteria produced was soaked up by other molecules dissolved in seawater, especially iron. Iron acts like a sponge for oxygen. Without oxygen, iron can float free in seawater. However, oxygen combines with free iron to form the ferric oxide (rust) that we all know too well. As we can now see in the geological record, iron and oxygen settled out into huge bands of reddish sediment, and the levels of dissolved iron in the ocean plummeted. Once all the dissolved iron was used up, the amount of free oxygen in the ocean began to rise. When the ocean could no longer hold all the oxygen produced by the photosynthetic newcomers, it began to spill out into the atmosphere.

Today, earth's atmosphere is around 21 percent oxygen with almost all the rest being nitrogen. Before the advent of the photosynthetic bacteria, atmospheric oxygen levels were close to zero; by about 2.4 billion years ago, they were rising rapidly to levels unseen since the birth of the planet, though barely reaching 1 percent of what we measure in the air today. This is referred to as the Great Oxidation Event, and it led to major changes in the environment. For one, as oxygen accumulated in the atmosphere, it eventually began to form ozone (O_3) at higher levels. This ozone layer served as a shield to reduce the penetration of DNA-damaging ultraviolet (UV) radiation from the sun, and eventually made it possible for organisms to live on land.

To this point in earth's history, we've seen four major developments: the first appearance of cellular life in the form of bacteria; the generation of a bigger and badder kind of bacteria called archaea; the evolution of photosynthetic bacteria; and the initial sign of atmospheric oxygen. But what is perhaps the most important innovation in the history of life was still around the corner.

Enter the Eukaryotes
. . . .

Somewhere around two billion years ago, shortly after the Great Oxidation Event, a new form of life appeared on our planet, and it was unlike anything the earth had ever seen before. It was a type of cell so different that it was given the name *eukaryote*, which means "true nucleus." These new cells were thousands of times bigger than bacteria and archaea. Bacteria and archaea were relatively peaceful species. We can think of them as vegans—consuming only small "natural" molecules like sugars or even making their own using simpler building blocks such as carbon dioxide and hydrogen gas. In contrast, eukaryotes were hunters. They developed the ability to eat (phagocytize) other cells such as bacteria and archaea. More than anything, they were able to thrive in the new oxygen-rich environment. Within a few million years, they had spread all over the planet.

So how did this incredible new life form come about? This topic was hotly debated for many years. We can think of pro-karyotes (bacteria and archaea) as primitive cells. They have no nuclear membrane surrounding their DNA, so their genetic material floats free within their cytoplasm—the liquid interior of the cell. What has become clear over the last few years is that eukaryotes arose from a one-in-a-million union between an archaeon and a bacterium. We don't know how, and we don't know why, but a little over two billion years ago, a bacterium invaded an archaeon and they decided to make it a permanent arrangement. The result was the first eukaryote.

Now, this never should have happened. As we'll learn later, both archaea and bacteria produce energy by means of protein machines embedded in their membranes. Because of their small size and the way their membranes are studded with thousands of these little protein energy generators, there's really no way for them to swallow an organism of similar size. Perhaps the relationship started as a close-buddy system rather than a symbiosis. After all, the new photosynthetic bacteria had learned to deal with the toxic effects of oxygen—they had figured out how to get close to the fire without getting burned. As oxygen levels began to rise in the oceans, archaea that otherwise would have been poisoned by this corrosive gas might have begun to cozy up to the new types of bacteria that had learned how to use oxygen to make energy. Imagine being able to walk through a room of sulfuric acid by covering yourself with insects that could breathe that gas and turn it into something harmless. Some archaea might have learned a similar trick. This could have allowed them to venture into areas with higher and higher concentrations of oxygen. Perhaps the archaea attracted the oxygen-processing bacteria by producing some sort of food molecule the bacteria needed. Over the course of millions of years, this relationship continued to tighten to the point that the bacteria took up full-time residence inside the archaeal cell.

If this all seems like a fairy tale, let me assure you that it's not. This symbiotic theory of the development of eukaryotes was all but proved when it was discovered that the remnants of these swallowed bacteria—the mitochondria—still have their own DNA more than two billion years after the initial merger. More on that topic later, but suffice to say that there is no doubt among scientists that eukaryotic cells arose from the unlikely union of an oxygen-processing bacterium and an archaeon.

And so we arrive at a point two billion years ago. The seas are full of bacteria, slightly tougher archaea, and the relatively giant eukaryotes—all single-celled organisms trying to survive in various ecosystems around the planet. Life was strangely quiet for the next billion years. In fact, the period between one and two billion years ago is often called the "boring billion." We can see traces of all three different types of unicellular life in the fossil record, but nothing else. No plants. No animals. However, life was getting ready to exploit the next big breakthrough: multicellularity.

The Rise of the Eukaryotic Federation

It may seem strange that the primitive prokaryotic cells—bacteria and archaea—never figured out how to group together into specialized federations. They came close. Some bacteria live in various sheets and clumps, and many exist as part of a larger eco-system. But none of them took the concept any further. Starting roughly one billion years ago, however, eukaryotes began to realize that they were stronger together than apart, at least in some cases. There are still types of eukaryotic loners that prefer to go their own way, like outlaw cowboys in old Western movies. We call these single-celled eukaryotes *protists*. Amoebae fall into this category. However, about a billion years ago, some types of eukaryotes began to live together permanently in what we might call a body.

These early multicellular eukaryotes were very simple, but over time different groups of cells started to specialize in performing different functions. This is similar to what we see in advanced societies made up of people who are more or less interchangeable but who contribute in different ways to the overall well-being: doctors, nurses, firefighters, soldiers, teachers, etc. Jellyfish are an example of early multicellular eukaryotes. About a half-billion years ago, something set off an explosion in the types and varieties of multicellular creatures, some consisting of billions or even trillions of eukaryotic cells living together in permanent union.

The advantage of this approach is obvious—it allows different cells to specialize in a way that makes the total assembly more powerful. Why this innovation had to await the development of eukaryotic cells is less obvious, and why it took over a billion years is even harder to explain. It may well be that the climate on earth wasn't favorable earlier. Perhaps there was insufficient oxygen or too much of the planet was covered by glaciers. Regardless, the fossil record is clear that by a half-billion years ago, large multicellular plants and animals thrived all over the land, seas, and skies with all the body plans we're familiar with today: notably, bilateral symmetry and three different layers of cells—the endoderm (stomach, intestines, etc.), mesoderm (muscles, bones, etc.), and ectoderm (skin, brain, and spinal cord).

The period that started about a half-billion years ago is called the Cambrian Explosion. During this time, all sorts of new complex animals began to appear in the fossil record. These included dragonflies with wingspans of three feet, huge worms that crawled over the earth, and many other large creatures that would be terrifying to encounter in a dark alley. Oxygen levels had evidently risen to exceed those of today, so the evolution of these huge beasts became possible. All the body plans we currently recognize in animals were born during this period.

The Cambrian Explosion came to a dramatic close a couple of hundred million years later with an event known as the Permian Extinction. About a quarter-billion years ago, conditions on our planet changed, probably because of sporadic periods of massive glaciation. This environmental shift caused the majority of existing species to die out completely, never to walk the earth again. The Permian period lasted from about 250 million years ago to sixty-six million years ago and became known as the age of the dinosaurs. These huge reptiles dominated the planet for over a hundred million years until a great cataclysm wiped them out in the blink of an eye. Most scientists believe this sudden extinction was caused by a massive meteorite that hit just off the Atlantic coast of Mexico, kicking up ash and dust that quickly darkened the sky around most of the globe. The temperature of the earth dropped, and many of the plants favored by the dinosaurs died out. With their primary food source depleted, the dinosaurs themselves rapidly went extinct.

The age of the dinosaurs saw not only the proliferation of these reptilian behemoths but the evolutionary arrival of a strange new type of animal—the mammal. The first mammals were small and hid from the many predators during the day, coming out to forage for food at night. They brought many innovations such as the ability to maintain their internal temperature so they could adapt to

Fig. 5: Timeline

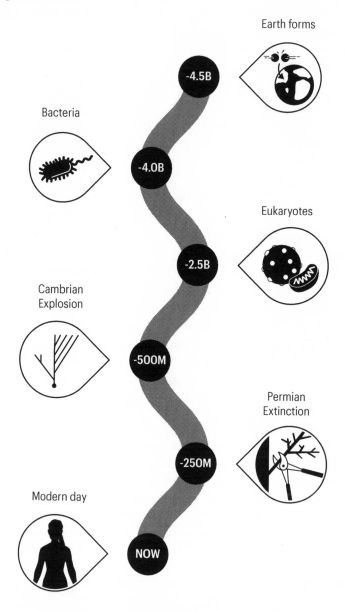

just about any habitat. They also nurtured their young far longer than had been the case for other types of animals. This allowed newly born mammals to start life with more complex brains that could be used to learn and adapt rather than coming into the world with most behavior hardwired. When the giant meteor wiped out the dinosaurs and countless other large species, mammals rushed in to fill the ecological void. Their small size allowed them to survive on less food, and thermoregulation allowed them to inhabit a variety of ecosystems. And their relatively large brains were well suited to a rapidly changing environment in which the right decision made the difference between life and death.

To finish off our overview of the history of life, we'll quickly cover the most advanced mammals of all. Primates appeared about fifty million years ago, just a dozen or so million years after the dinosaurs became extinct. These new mammals had even larger brains and an even greater capacity for social interaction. Of course, one branch of primates eventually led to the Homo line, of which we are descendants. The Homo genus probably evolved ten to twenty million years ago and went through a series of fits and starts until *Homo sapiens* first appeared in Africa roughly two million years ago. The rest, as they say, is history.

References and Further Reading

D Baker, T Wijshake, T Tchkonia et al (2011). Clearance of p16Ink4a-positive senescent cells delays aging-associated disorders. *Nature*, 479: 232–36.

HC Betts et al (2018). Integrated genomic and fossil evidence illuminates life's early evolution and eukaryote origin. *Nature Ecology and Evolution*, 2: 1556–62.

2

A Crash Course
in Chemistry

"Chemistry begins in the stars. The stars are the
source of the chemical elements, which are the building
blocks of matter and the core of our subject."

PETER ATKINS

I N THE previous chapter, we talked about how we can run the
model of evolution backward to imagine how life might have
developed. We can do the same thing with models of physics
to see how the universe was formed. When we do, all the lines
converge at a strange point about 13.8 billion years ago when all
the matter in the universe was so concentrated that it could fit in
the palm of your hand. Even the most advanced simulations can't
tell us how or why. From there, it looks as if the universe suddenly expanded, sending streams of mass and energy off in every
direction. Over the course of a few million years, some of this
mass settled into the tiny particles that make up atoms. For the
purposes of this book, we care only about two kinds of particles—
protons and electrons. You might recall from high school that

protons contain a positive charge and lie at the center of atoms. We call this central area the nucleus. Whirling around it just as the planets circle the sun are the negatively charged electrons. That's actually a vast oversimplification, but let's run with it.

At first, the only type of atom in the universe was the simplest— hydrogen. It consists of a single proton with one electron in orbit around it.

Fig. 6: Hydrogen atom

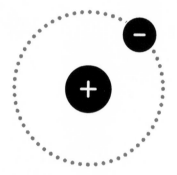

At a few points in space, hydrogen atoms began to clump together, eventually forming massive stars. As these gas balls blazed in the heavens, heavier elements began to form in their interiors. First up was helium—essentially a pair of hydrogen atoms combining to contain two protons and two electrons. If you think you see a pattern, you're right. Normally, the number of protons equals the number of electrons in an atom to keep the net charge at zero.

Fig. 7: Helium atom

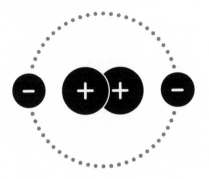

This gives us the first row in the periodic table. Why are there only two types of atoms (or "elements") in this row? Because electrons orbit the nucleus in different "shells," just like the planets orbit the sun at various distances (Mercury, then Venus, then Earth, etc.). Each row of the periodic table corresponds to one of these electron shells. Unlike the solar system, however, each orbit (shell) can hold a specific number of electrons. The first shell can hold up to two electrons, and thus the first row in the table has hydrogen with one electron and helium with two.

Fig. 8: First row of periodic table

Later in their lives, the first stars began to form even heavier elements. Hydrogen and helium atoms collided to form lithium with three protons and three electrons. Other fusions began to form the rest of the elements in the second row of the periodic table. This row contains eight elements. Can you guess why?

That's right, because the second electron shell can hold eight more electrons. So lithium (symbol Li, atomic number 3) looks like this.

Fig. 9: Lithium atom

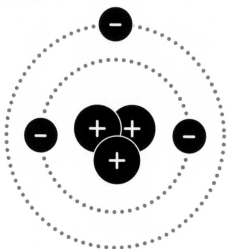

As you can see, a lithium atom has three protons and three electrons. Two of the electrons go into the first shell while the third electron goes into the second shell. Since the second shell has only one electron, this puts lithium in the first (leftmost) position in the second row of the periodic table.

Fig. 10: Periodic table with lithium

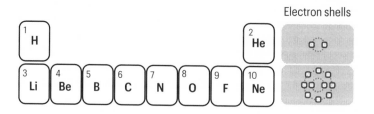

In figures 8 and 10, the boxes on the right of the first two lines of the periodic table indicate the available spaces for electrons in the first and second shell.

I PROMISE, we're not going to spend a lot more time on chemistry, but there are a few additional concepts that are important to understand. When an atom's outermost shell is full, it is pretty content. Helium and neon (symbol Ne, atomic number 10) fall into this category. Helium has two electrons in its outer shell. Since this is the first shell and holds only two electrons, helium is very unreactive. Neon has ten electrons to match its ten protons: two in its first shell and eight in its second. Since the second shell has capacity for only eight electrons, neon is fat and happy too. Like helium, it tends to be relatively unreactive.

Carbon, the Element of Life

Now let's focus on the most important element of all, at least as it pertains to life—carbon. Why is carbon so important? Why do we talk about life as being carbon-based? To answer these questions, can you figure out how many electrons carbon has in its outermost shell? It's pretty easy. Just count to the right starting at lithium. We know that lithium atoms contain one electron in their outermost (second) shell. This means that beryllium (Be) contains two and boron (B) has three. So carbon has to have four electrons in its outer shell, as illustrated in figure 11.

Fig. 11: Carbon atom

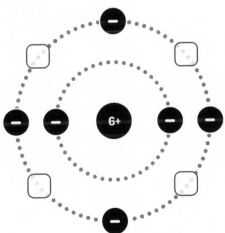

In figure 11, we can see that carbon has four empty spots in its outermost shell. These are the spots available for creating bonds with other atoms. This means that carbon can form up to four additional bonds. In a nutshell, this is what is so special about carbon. It is the smallest and most plentiful element capable of bonding with four other atoms. In contrast, consider the case of oxygen, shown in figure 12, two positions to the right in the same row.

Fig. 12: Oxygen atom

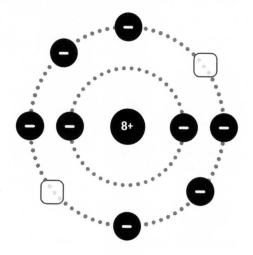

Oxygen has six electrons in its outermost shell, leaving two positions free. This means that oxygen can form two bonds at most.

Please don't shudder at the question, but what exactly is a chemical bond? At the most superficial level, it's really quite simple. A chemical bond occurs when two atoms decide to share a pair of electrons. The simplest example is hydrogen gas, shown in figure 13.

Fig. 13: Hydrogen gas molecule

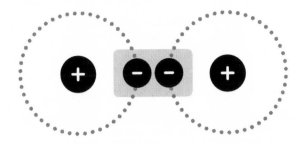

A *molecule* is two or more atoms connected by this sort of electron-sharing bond, which is given the name *covalent*. So a molecule of hydrogen gas consists of two hydrogen atoms that have decided to share their solo electrons. In this way, both hydrogen atoms "feel" (to put it simply) like they have two electrons—the one they each started with and the one they are sharing with their neighbor. The physical analogy would be two people grasping arms. Yes, it's possible to break them apart, but only with a lot of effort.

Fig. 14: Grasping arms

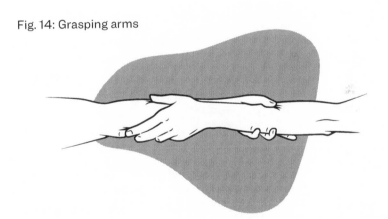

Let's consider the example of another common molecule—water. As you know, the chemical formula for this familiar substance is H_2O, two atoms of hydrogen and one molecule of oxygen. These three atoms are arranged as shown in figure 15.

Fig. 15: Water molecule

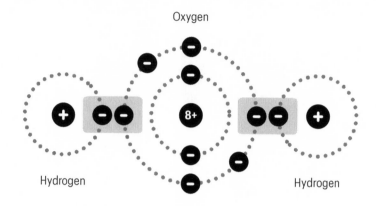

So each hydrogen atom is bonded to the lone atom of oxygen. This makes the oxygen atom feel like it has eight electrons in its outer shell—the six it started with and one from each of the hydrogens. Each hydrogen shares one of the oxygen atom's electrons, making it feel like it has a full complement of electrons in its outer shell. So everyone is happy. Oxygen effectively has a full outer shell of eight electrons and each hydrogen effectively has two electrons in its outer shell. (Please note that the actual three-dimensional shape of water is slightly different—rather than being arranged in a straight line as shown in figure 15, the hydrogen atoms actually connect in more of a V shape with the oxygen at the bottom. We'll come back to this point later.)

Let's return to our favorite type of atom—carbon. As mentioned, it's special because it is the smallest atom able to form four bonds because of the four electrons and four free spots in its outer electron shell. In contrast, hydrogen is pretty boring. It can bond to only one other atom at a time. Oxygen can usually form, at most, two bonds. Only carbon can create the variety of shapes we see in nature—all sorts of complex chains, circles, and three-dimensional networks. Let's look at a few of these. One of the simplest is methane, in which a central carbon atom is bonded to four hydrogen atoms.

Fig. 16: Methane

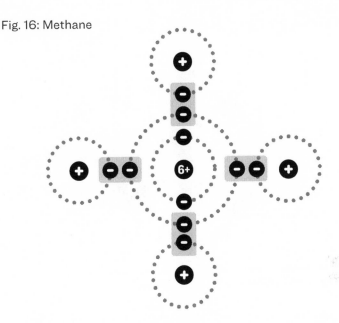

As in the case of water, each individual atom is happy. Because carbon is sharing each hydrogen atom's lone electron, it feels like it has the full complement of eight electrons in its outermost shell. Remember that carbon started with four of its own electrons in its second shell. Similarly, each hydrogen gets to share one of those four original carbon electrons and thus feels like it has a full outer shell with two electrons. However, since hydrogen can form only one bond, there's no room for any additional growth.

The situation starts to get more interesting when we bind carbon atoms to other types of atoms—especially to carbon itself. What happens if each carbon binds to four other carbons, and each of those binds to four additional carbons, and so on? That forms the crystalline structure we call *diamond*—the hardest natural substance we know.

Fig. 17: Diamond

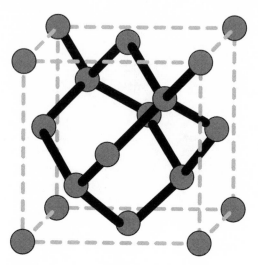

In the diagram of diamond's molecular structure, each ball represents a carbon atom, and the sticks represent two electrons forming bonds between pairs of carbons. Another all-carbon crystal arrangement can be found in *graphite*. Here, each carbon is bonded to only three other carbons all in the same plane. This allows each "sheet" of graphite to easily slide off, making it a very soft substance.

Fig. 18: Graphite

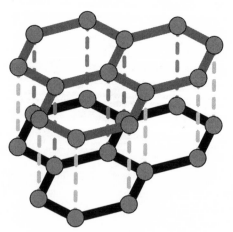

Philanthropists and Misers
....

So far, we've treated chemical bonds as mutually beneficial arrangements in which two atoms share electrons—like a friendly divorce in which the parents have equal custody of the children. However, some types of atoms are more "electron greedy" than others. When a miserly, electron-greedy atom forms a bond with a more philanthropic atom, the resulting bond is skewed. Instead of spending half their time with both parental atoms, the two electron children in the bond tend to be monopolized by Mom or Dad. The classic villain in this scenario is oxygen. Think of oxygen as a miserly bully. It wants both electrons in a bond all to itself. We can see this clearly in a molecule of water, as shown in figure 19.

Fig. 19: Water molecule

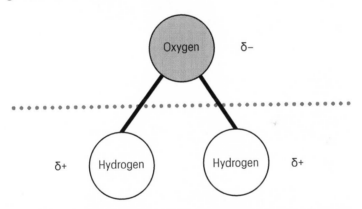

In this illustration, the lines between the hydrogen and oxygen atoms are symmetrical, but you can imagine the oxygen atom pulling the two electrons in each bond closer to itself. This gives the oxygen a partial negative charge (represented by δ-) and leaves each hydrogen atom with a partial positive charge (δ+). Oxygen's hunger for electrons lies at the core of most of life's energy production. Compared to oxygen, other elements such as hydrogen and carbon are pushovers. They are all too happy to participate in an uneven electron-sharing arrangement with oxygen. Compared to miserly oxygen, they are relatively philanthropic, willing to let oxygen monopolize or even steal their electrons outright. It's this uneven desire for electrons that produces fire.

Let's consider the case of methane. We can redraw methane with each line representing the two electrons forming a bond. Remember that in this particular case, one of the electrons in each bond comes from the outer shell of the carbon atom, while one comes from the outer shell of a hydrogen atom.

Fig. 20: Methane

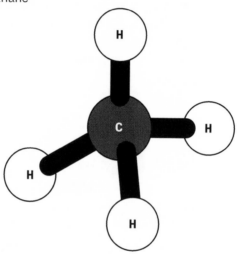

Now imagine that oxygen comes onto the scene. Our most common encounter with this gas is in the air, which is about 21 percent oxygen. Here oxygen doesn't exist as single atoms but rather as molecules, each containing two oxygen atoms linked together by covalent (electron-sharing) bonds.

Even though each oxygen atom feels like it has the full complement of eight electrons in its outer shell, neither is truly happy. Remember that oxygen is an electron miser. The two oxygen atoms are identical and unable to bully each other. Just like a bully in real life, they would prefer to find a weakling they can dominate. Methane is like a skinny nerd with money in his pocket—a perfect target for oxygen. The currency that methane has and that oxygen wants is electrons.

Fig. 21: O_2 molecule

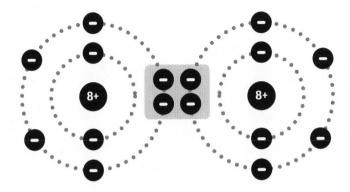

Under the appropriate circumstances, two oxygen molecules will gang up on poor methane and steal most of its electrons. The chemical formula for the process looks like this:

CH_4 (methane) $+ 2O_2 \rightarrow CO_2 + 2H_2O$

We can imagine one of the oxygen molecules (two oxygen atoms) jumping methane in a dark alley and ripping off the four hydrogen atoms that were previously sharing electrons with the carbon atom. Each oxygen atom makes an offer to two hydrogen atoms that they can't refuse—"Instead of sharing that electron with carbon, share with me. Or else." This yields two water molecules, as in this chemical formula:

$O_2 + 4H \rightarrow 2H_2O$

This mugging leaves the poor carbon atom with just four electrons in its outer shell and no one to share them with. The second oxygen molecule (a pair of oxygen atoms) rushes in with an offer phrased something like this: "Hey, sorry to see that you're now four electrons short of a full outer shell. Poor thing. What if we two oxygen atoms share ours with you?" This combination results in the following chemical equation:

Carbon with four electrons (Carbon Lewis structure)

$$\cdot \overset{\displaystyle \cdot}{\underset{\displaystyle \cdot}{C}} \cdot + O_2 \rightarrow CO_2$$

It turns out that the amount of energy contained in the original methane molecule and the two oxygen molecules is greater than that in the two water molecules and the single carbon dioxide molecule that result from this burning. We know that energy can't just disappear, so where does it go? Well, the complete picture could be expressed like this:

CH_4 (methane) $+ 2O_2 \rightarrow CO_2 + 2H_2O +$ Heat + Light

So burning methane by allowing it to react with oxygen generates not only water and carbon dioxide but also heat and light—exactly what we feel and see around a fire! Another name for methane is natural gas; you make use of this reaction every time you use a gas stove.

If we step back and look at the big picture, what we see is this: Fuel molecules are quite often carbon atoms with lots of hydrogen atoms attached—so-called hydrocarbons. When they react with oxygen, two things happen:

1 Oxygen atoms can steal hydrogen atoms, including the electron the hydrogen atom was previously sharing with carbon in the fuel molecule.

2 The remaining carbon atoms, stripped of their hydrogens, combine with oxygen to form carbon dioxide.

And the more hydrogen atoms a fuel molecule has, the more energy can be produced by combining it with oxygen. This is the process we call *combustion*. Methane is a highly flammable gas that is sometimes found in large deposits in the ground and pumped out for energy production. What follows are some other powerful hydrocarbons that can serve as fuel.

Ethane has two carbons bound together, with each carbon atom also bound to three hydrogen atoms.

Fig. 22: Ethane

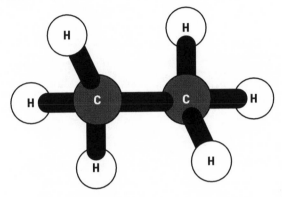

Octane is a chain of eight carbon atoms with all the other available bonds taken up by hydrogen atoms. It is the most important constituent of gasoline.

Fig. 23: Octane

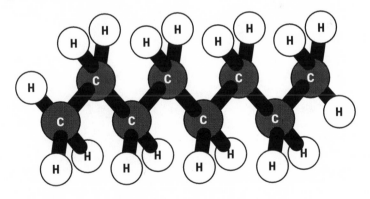

We can conclude that strings of carbon molecules with their other available bonds connected to hydrogen molecules are highly flammable fuel molecules. This is because they contain lots of electrons and hydrogen atoms for oxygen to steal, generating heat and light in the process, just as in the combustion of methane. So, when you see hydrogen atoms attached to carbon atoms, think "That's energy!"

Let's test your knowledge by examining the following one-carbon molecules:

Name	Chemical Formula	Structural Diagram
Methane	CH_4	
Methanol	CH_3OH	
Formaldehyde	CH_2O	

Name	Chemical Formula	Structural Diagram
Formic Acid	CHO_2H	
Carbon Monoxide	CO	
Carbon Dioxide	CO_2	

Which of these molecules contains the most usable energy for combustion? Which one comes in second? Which one has the least?

And the answer is... Methane has the most usable energy because of its four hydrogens bonded to carbon. Methanol also has four hydrogens, but only three of them are bound to carbon. The one bound to oxygen is less useful in terms of energy production.

At the other end of the spectrum, you might say that both carbon monoxide (CO) and carbon dioxide (CO_2) are tied for least usable energy because neither has any hydrogen bonding partners. However, remember how much of an electron bully oxygen is. In carbon monoxide, the carbon atom is triple bonded to oxygen, sharing six electrons between them. For reasons we won't go into, two of those electrons are supplied by carbon and four by oxygen. The two electrons shared by carbon are essentially lost because they are monopolized by oxygen. This effectively leaves the carbon with two unshared electrons. In carbon dioxide, the carbon shares two of its electrons with one oxygen atom and two with another, leaving it with no unshared electrons. Combustion is actually more about electrons than hydrogen atoms, and carbon monoxide has one less electron that it can give up under the right conditions. So we can say that the most oxidized of these molecules is carbon dioxide, with carbon monoxide a close second. You can think of oxidation as the process of having your lunch money (electrons) stolen by the school bully—oxygen. The CO_2 molecule has been completely victimized by oxygen. It has nothing left to give.

A Glimpse Ahead

But why even bother with all this talk about electrons? Well, it turns out that metabolism is all about them. You might have heard that ATP (adenosine triphosphate) is the energy currency of the cell. That's not quite true. ATP is more like a debit card, and the dollars come in the form of electrons. So the big picture is this:

1 You can extract energy from carbon-based molecules by stealing their electrons, especially the ones shared with carbon by hydrogen atoms. This frequently involves the electron bully, oxygen.

2 You can fund construction projects by spending (or donating) electrons. All the big construction projects in the cell (making proteins, DNA, etc.) depend on acquiring electrons from somewhere.

We'll be talking much more about how these electrons are handled throughout the cell in the chapters ahead.

References and Further Reading

P Ball (2000). *H2O: A Biography of Water*. Orion Publishing.
P Levi (1975). *The Periodic Table*. Schocken Books.

3

Life 1.0

S O FAR, we've taken a bird's-eye view of the formation of the earth and the timeline of the development of different forms of life on it. We've also had a crash course in chemistry and now understand the basic chemical principles underlying life. We know that the universe began with a single type of atom—hydrogen—that has one lonely proton orbited by a single electron. The element helium was born in the center of stars when extreme temperatures and pressures caused pairs of hydrogen atoms to fuse together, resulting in a new type of atom with two protons orbited by two electrons. This fusion process continued and created the dozens of other types of atoms (or elements) we see in the periodic table.

The elements we particularly care about are carbon and oxygen. As we learned, carbon is special because it is the smallest atom (six protons) that can form bonds with four other partners. This allows carbon-based molecules to form a dizzying array of shapes, including chains, rings, sheets, and even crystals. Recall that bonds are simply pairs of electrons shared by two atoms. And finally, remember that oxygen is an electron miser. It will steal electrons from weaker atoms outright. And when it "shares" electrons with other atoms, oxygen doesn't do so equally—it only grudgingly allows shared electrons to spend time with other atomic partners. This tendency of oxygen to hang on to electrons is the basis for the metabolism that we'll be discussing in the rest of this book.

Now let's go back to the beginning—about four billion years ago—and look more closely at the origin of those very first bacterial cells. How did life get started in the first place?

As we've noted, the early earth was extremely inhospitable. It was a searing hot place dominated by violent volcanic activity. For millions of years after the planet was formed, it was bombarded by huge chunks of rocky material left over from the formation of the solar system. In fact, the prevailing theory on the origin of the moon is that it was the product of a collision between the early earth and another would-be planet in the same orbit, called Thea. This massive impact incinerated Thea and knocked a big piece of the earth out into an orbit just a few thousand miles away, much closer than it is today.

The early moon would have been an incredible sight to behold from the surface of the earth. It would have dominated the sky and appeared many times larger than the sun. Because it was so much closer to earth, its gravitational pull created huge tides: first of molten magma before the earth cooled, and later of water once the oceans formed. Imagine waves hundreds of feet high occurring frequently because the earth was rotating on its axis every five or six hours. Gradually, the moon moved out to its current distance of about a quarter of a million miles away, and the earth's rotation slowed to its present twenty-four hours.

By about four billion years ago, things had started to settle down. The earth had cooled enough for liquid water to begin to form from the steam emitted by volcanos and the ice delivered by raining meteors. The rocky debris from the formation of the solar system had almost all been pulled into the sun or planets by gravitational forces, so it struck the earth much less frequently. Conditions were now favorable for the development of life.

How Life Began

Life. We use that word so casually… but what exactly *is* life? If you've never pondered the question before, it might seem silly. Life is living things, you might say—plants, animals, mushrooms, and monkeys. Life is self-evident. Even infants can tell the difference between living things and inanimate objects. Give a child a kitten and a teddy bear, even one that moves and talks, and see which elicits more squeals of delight. The child can immediately tell which one is alive, even from a very early age.

We could spend a long time trying to define life, but let's just make a quick pass over the subject for brevity's sake. At least on our planet, all living things are comprised of cells. Said another way, a cell is the smallest thing that can be considered alive. All cells are made up of the same basic components:

- One or more sets of long DNA molecules that contain recipes for proteins. For example, one section of DNA in your cells contains the recipe for hemoglobin—an important protein used by red blood cells to transport oxygen around your body.

- Machines called ribosomes that take the recipe for a protein and manufacture it. DNA also contains the recipe for making the ribosomes themselves.

- Lots of proteins that carry out the business of the cell. Some proteins move things around within the cell. Others act as doorways to allow nutrients and other desirable substances into the cell while keeping most substances out. As we'll see, proteins are also responsible for helping the cell generate the energy it needs to stay alive.

* A fatty membrane that surrounds the cell and acts as a wall to keep the contents of the cell in and most of the outside world out. This fatty wall is referred to as the *cell membrane* or *plasma membrane.*

The ultimate goal of the cell is to make copies of itself, although the situation is more nuanced in multicellular plants and animals. In this chapter, we're going to concentrate on the very first and most primitive cells—bacteria.

Even though bacteria are dramatically simpler than the eukaryotic cells that make up our bodies, they are still mind-bogglingly complex. In particular, in the bewildering variety of ways bacteria are able to generate energy from just about any substance you can imagine. Given this complexity, scientists have struggled to understand how they came to arise on the early earth. We seem to constantly run up against the chicken and egg problem. Consider these somewhat contradictory facts:

* All cells today rely on proteins to do most of their work, whether it's generating energy, making copies of DNA, or constructing new proteins.

* DNA is required to provide the recipe for proteins.

* Cells need membranes to separate themselves from the outside world. But the membranes of today's cells are studded with proteins that act as selective channels that let in only things the cell needs and get rid of waste.

So which came first—DNA or proteins? If DNA came first, how did it get constructed and copied without proteins? If proteins came first, how were they made without a recipe stored in DNA? How did membranes come into the picture? And how did they learn to incorporate proteins to serve as border guards?

Entire books have been written about possible origins of life. And the subject engenders lively debate among scientists to this day. This book broaches this subject as a means to an end—to help understand the function of the body and the mechanisms underlying health, disease, and aging. So we'll ignore the many controversies and alternative theories in favor of a simple "just so" story to help give the big picture; but please realize that this is a rapidly developing field, and conclusions are likely to change in the years to come.

The whole field was set abuzz with excitement back in the 1950s with the famous Miller–Urey experiment. Stanley Miller was a graduate student in Harold Urey's lab at the University of Chicago. In his celebrated experiment, Miller set up a contraption containing a gas mixture similar to that thought to have existed on the ancient earth—hydrogen, methane, and ammonia. He then used a periodic electrical current to simulate lightning strikes. To the astonishment of the researchers and the entire scientific world, the experiment quickly generated many of the building blocks of life: amino acids, lipids, carbohydrates, and nucleotides (components of DNA). This led many to believe that life originated in a "primordial soup" in lakes and oceans on the early earth. But this theory suffered from one major flaw—in this watery soup, the building blocks would be incredibly dilute. How could they possibly come together to form the more complex assemblies of proteins, DNA, and RNA required for life?

The Alkaline Vent Hypothesis

A new theory arose in the 1980s and still holds sway today—that life originated in undersea hydrothermal vents. You see, the ocean floor is periodically broken by underlying deposits of hot liquids and gases. At these vents, boiling water streams up from deep within the earth. Some of these vents are alkaline—that is, the opposite of acidic—and somewhat cooler. And the water coming from these vents contains lots of electron-rich gases: molecular hydrogen (H_2) and others. This vent water is much different from the overlying ocean water, which is colder and more acidic with lots of dissolved carbon dioxide (CO_2). This was especially true when life first arose approximately four billion years ago.

Mineral deposits around these alkaline vents created countless small pores through which the vent water would circulate before reaching the open ocean. Interestingly, these pores were similar in size to bacterial cells—roughly one micron in diameter. Could it be that life originated in rocky minerals, using their thin pores as enclosures instead of the fatty membranes that cells use today? The concept seems increasingly plausible and solves some of the chicken and egg problems posed above.

The alkaline vent hypothesis provides an elegant answer to the question, "Which came first, metabolism or heredity?" As mentioned earlier, in modern cells, both metabolism and heredity require proteins. But the construction of proteins requires both energy from metabolism as well as a heritable protein recipe stored in DNA. In alkaline vents, metabolism might have come for free, eliminating the need for either proteins or genetic material, at least at first.

Fig. 24: Hydroelectric power generation

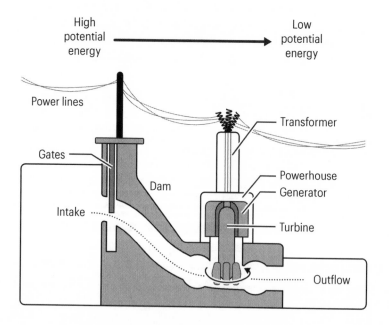

To understand how metabolism might have originated, let's first consider how modern bacteria generate energy. Do you know how dams produce electricity? Consider figure 25, a diagram of hydroelectric power generation.

The "secret sauce" is the way that water moves from a higher place to a lower one. The force of the water flowing downhill turns the blades of a turbine. This turning motion generates electricity. The Hoover Dam on the Arizona–Nevada border produces enough electricity to cover the needs of more than a million people each year. People discovered the latent power in falling water long ago. Consider this centuries-old mill powered by falling water.

Fig. 25: Water-powered mill

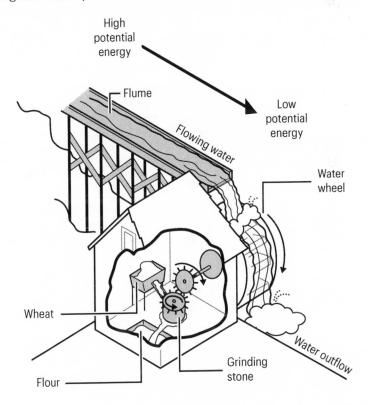

In this case, the flume on top delivers water from a higher source such as a lake or uphill stream. The falling water hits the blades on the water wheel, causing it to turn. The wheel is connected to a shaft used to rotate a large grinding stone that can wear down wheat or other grains to produce flour, as shown in the cutaway.

What is the ultimate source of this power? Gravity. We say that the uphill water has more *potential energy* than the downhill water, just as a bowling ball in your hand has more potential energy than the same bowling ball on the floor. If you drop a bowling ball onto your foot, gravity turns this potential energy into motion (kinetic) energy and you quickly feel the result.

It may surprise you to learn that cells employ more or less this same principle—using potential energy to generate power. But instead of tapping into water flowing downhill, cells do something very similar with protons.

Fig. 26: ATP-synthase & proton gradient

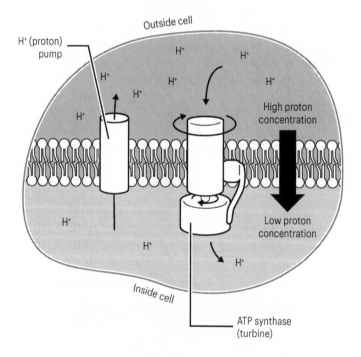

You'll notice that the top of figure 26 represents the outside of a cell, while the bottom represents the inside. Between them is a bilayer (double layer) membrane of lipid molecules (fat). Remember that a hydrogen atom consists of one proton and one electron. If we strip the electron away, just the proton remains. This is often drawn as H^+. Notice how the outside of the cell has lots of protons while the inside has relatively few. This difference in concentration is just like the difference between water uphill and water downhill—it represents potential energy. One of the most amazing machines in the universe is a set of proteins called ATP synthase, inserted into the center of the membrane (see figure 26). Just as falling water can turn a turbine or a water wheel to generate power, protons flowing from an area of high concentration (the outside of the cell) to an area of low concentration (inside the cell) turn the "blades" on ATP synthase. This turning motion is used to squeeze a phosphate onto a molecule called ADP (adenosine diphosphate) to create a molecule of ATP (adenosine *tri*phosphate).

As you may recall from high school, ATP is the universal energy currency of the cell. You produce roughly your body weight in ATP every day. ATP is used to power countless reactions within the cell to help you grow and live. And you can see that there's one more piece to this puzzle represented by the cylinder inside the left side of the membrane. This is another tiny protein machine—a pump that takes protons from inside the cell and pushes them out, thus maintaining the difference in concentration (or *gradient*). Of course, this pump requires energy to do its work. We'll cover this part later, but that energy can come from burning food in the case of most cells, or from sunlight for photosynthetic organisms.

It's all well and good that ancient cells had these wonderful little protein machines and honed them over billions of years for peak efficiency. But doesn't that just make it even harder to explain how life began? There's no way that primitive cells could have synthesized such complex devices early on. The great thing about the alkaline vent hypothesis is that they didn't need to. There was already a proton gradient in place for these primordial cells to exploit, as you can see in figure 27, a diagram of an inorganic pore. This eliminates the need for a pump to maintain the difference in H+ concentrations.

Fig. 27: Inorganic pore

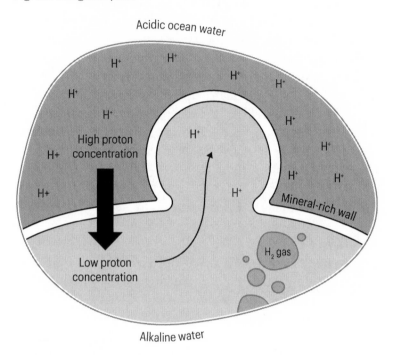

The early ocean was acidic because it had high levels of dissolved carbon dioxide. Carbon dioxide reacts with water to create carbonic acid, which is why sodas can dissolve the enamel on your teeth. By definition, an acid releases protons (H$^+$). The chemical reaction is shown below.

$$CO_2 + H_2O \rightarrow H_2CO_3 \rightarrow H^+ + CO_3-$$

This is also one reason that high levels of CO_2 in the atmosphere are so dangerous—they acidify all water on the planet, even rain.

You can see from figure 27 that in alkaline vents, alkaline water (low in H+) enters a mineral pore where it is separated from the acidic (H+ rich) ocean water by a thin wall. The walls in these pores are often rich in minerals containing iron, nickel, and sulfur. As it turns out, these Fe(Ni)S minerals are able to catalyze (speed up) lots of different reactions, especially those that involve moving electrons from one molecule to another. In particular, these iron-sulfur clusters might have been able to perform the following magic tricks:

1 Add hydrogen atoms (with their electrons) to carbon dioxide to make more energy-rich molecules, especially acetate (CH_3COO-).

2 Squeeze acetate and phosphate molecules together to form acetyl phosphate—kind of a poor man's ATP.

If you look back to the previous chapter, containing the crash course in chemistry, you'll recall that carbon dioxide is the most oxidized or electron-poor single-carbon molecule. In addition to being alkaline, the water entering the pores from undersea vents is also rich in hydrogen gas (H_2). Iron-sulfur clusters are able to take the electrons from the hydrogen gas as well as one of its protons and push them onto carbon dioxide in a sort of "reverse burning" process:

$CO_2 \rightarrow CO \rightarrow CH_3OH$ (methanol)

From there, these same minerals can attach a carbon dioxide to methanol to produce the two-carbon molecule acetate. That's the same two-carbon molecule that forms the acetyl–CoA of high school biology fame.

Fig. 28: Acetyl-CoA

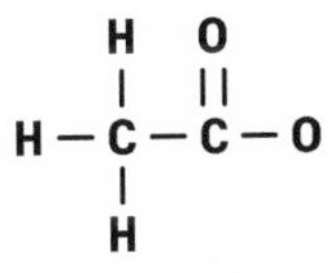

$$H - \overset{\overset{\displaystyle H}{|}}{\underset{\underset{\displaystyle H}{|}}{C}} - \overset{\overset{\displaystyle O}{||}}{C} - O$$

Acetyl-CoA

It's impossible to overstate the importance of these two processes. Once you're able to make acetyl phosphate from carbon dioxide, all sorts of doors open up. Acetyl molecules are the building blocks for just about everything in modern cells. They can be used to build fatty acids for membranes as well as the ribose sugar molecules that form the core of DNA and RNA. And the hydrogen atoms and their electrons can be stolen back from acetyl molecules to generate energy. As we'll see later, modern cells use acetyl molecules as gauges of their overall energy status and even attach them to proteins to turn those proteins on or off.

The Protocells Break Free

Now we have all the foundational components for early *protocells*— the forerunners to bacteria and other modern cells. All the energy they need is supplied by the earth itself, in the difference in temperature, acidity, and electron capacity between the water coming from undersea vents and the overlying acidic ocean. And mineral pores provide the encapsulation, allowing more complex molecules to be concentrated until they can ultimately link together to form amino acids and, from there, proteins. Eventually life could stumble on proteins that could make acetyl molecules and ATP even more efficiently than the iron-sulfur clusters present in these mineral pores.

But there's just one problem. Can you spot it? Because these first protocells relied on mineral pores to enclose them, they weren't exactly mobile. Their ability to divide was pretty limited too. The contents of a protocell could eventually spill over into an adjacent pore, providing a very primitive kind of replication. But that's far from the sort of free-living bacterial life teeming in today's oceans.

The encapsulation itself probably wasn't too hard to achieve. After all, modern cells make fatty acids by linking together acetyl groups. Fatty acids have a long hydrophobic (water hating) chain of carbons bonded to as many hydrogen atoms as will fit. And at one end of this chain is essentially a carbon dioxide atom.

Palmitic acid is the most common fatty acid in the membranes of cells today.

Fig. 29: Palmitic acid

Hydrophilic

```
        O  H  H  H  H  H  H  H  H  H  H  H  H  H  H
        ‖  |  |  |  |  |  |  |  |  |  |  |  |  |  |
H-O-C-C-C-C-C-C-C-C-C-C-C-C-C-C-C-H
        |  |  |  |  |  |  |  |  |  |  |  |  |  |  |
        H  H  H  H  H  H  H  H  H  H  H  H  H  H  H
```

Carboxyl group
(carbon dioxide)   Palmitic acid

The carbon dioxide (technically a carboxyl group), seen at the left end of the palmitic acid chain diagram, is hydrophilic (water loving) and dissolves well in water, so these molecules will self-assemble into layers like the one shown in figure 30, with the round part representing the hydrophilic end. To be precise, the circles actually represent another molecule (glycerol) to which two fatty acids and one phosphate molecule bind, but the overall principal is the same.

Fig. 30: Phospholipid self assembly

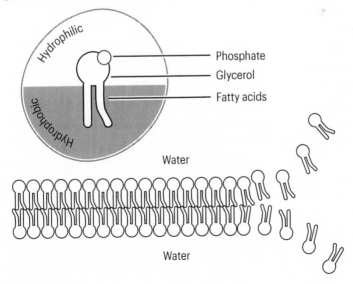

Hydrophilic

Hydrophobic

Phosphate
Glycerol
Fatty acids

Water

Water

It's easy to see how a protocell would have been able to manufacture enough of these fatty acids to enclose itself and float out into the vast ocean, eventually splitting in half to create more copies of itself. But by leaving the mineral pores near the vents, it would be abandoning its source of energy—the proton gradient across the wall of the pore. If we go back to our water wheel analogy, how could such an energy-generation approach be used to power a mobile device? It turns out that human engineers accomplished just this feat with the steamboat.

Fig. 31: Steamboat

Isn't it interesting how they took a stationary energy-producing scheme and adapted it to make a vessel move? The water wheel in a grinding mill made use of an existing energy resource (uphill water) and turned it into a new form (rotation of a wheel) that could do useful work (turn a stone to grind grain). Whoever first had the idea of powering a boat with such a water wheel must have asked the question, "But how do I replace the energy from gravity pulling water downhill to turn the blades?" This was eventually done with steam engines. A furnace on the boat was filled with wood or coal to make a fire. This fire was used to turn water into steam. The steam could then be used to rotate the water wheel and propel the boat.

The first cells came up with a similar solution. They still used protons flowing through a turbine to turn it and physical force to squeeze a phosphate molecule onto ADP to generate ATP. They even took some iron-sulfur clusters with them to help catalyze reactions. But by moving away from the vent pores, they were giving up the natural H+ gradient that had made life so easy. So they manufactured their own gradient by inventing a simple pump that could take protons from inside the cell and push them to the outside of the cell membrane. Think back to this diagram:

Fig. 32: ATP-synthase & proton gradient

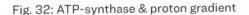

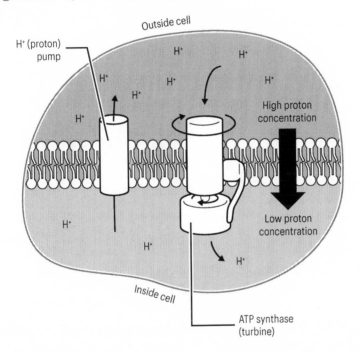

Now we have a nice little looping system. The pump on the left uses energy (we'll talk in a minute about where that energy comes from) to keep the H+ concentration higher on the outside of the cell than on the inside. The turbine-like protein machine in the center (ATP synthase) uses the energy of protons (H+) flowing downhill into the cell with its low concentration of H+ to turn the blades of the tiny turbine. This provides the force that can generate a molecule of ATP from a molecule of ADP and a molecule of phosphate. The energy to keep the pump running can come from several sources, but the most common is the burning of electron-rich food like sugar (glucose).

And so our little protocell is able to envelope itself in fatty acids, insert one protein pump into its membrane to maintain the proton gradient, and use that gradient to produce energy in the form of ATP. And it swims away from the vent system into the vast and murky ocean. There might have been several varieties of these early protocells, each taking slightly different approaches to generating energy or coming up with clever ways to store protein recipes and use them to build new proteins. However, biologists were amazed to discover that every single cell from every single type of life on earth adopted exactly the same approach for many critical decisions. In particular, the little factory (ribosome) used to produce proteins is almost identical across all branches of life. Other details are remarkably consistent: the three-letter genetic code used to specify a particular ingredient (amino acid) in a protein recipe (gene); the exact same twenty types of amino acids; and even the handedness (chirality) of those amino acids (every single one is left-handed). All this can lead to only one conclusion...

One Cell was the Last Universal Common Ancestor for All Life

It doesn't matter whether you consider bacteria, algae, plants, fungi, animals, or archaea—all have many, many features in common. Too many, in fact, for it to be a coincidence. The inescapable conclusion is that one type of protocell outcompeted all the others and continued to evolve into the dizzying variety of life-forms we find on our planet today. All cells store their genetic information in DNA using the exact same four chemical letters (A, T, C, and G). They all use the same three-letter sequence (e.g., CGG) to code for a particular amino acid (alanine) in their protein-coding genes. They use similar enzymes to make energy and repair damaged DNA. The list goes on and on. One type of protocell kept evolving—tweaking its energy production, DNA replication, cell division, import/export, and other machinery—until it became so efficient that it simply wiped out all its competitors. We call the resulting mature cell LUCA, for *last universal common ancestor*.

Fig. 33: LUCA

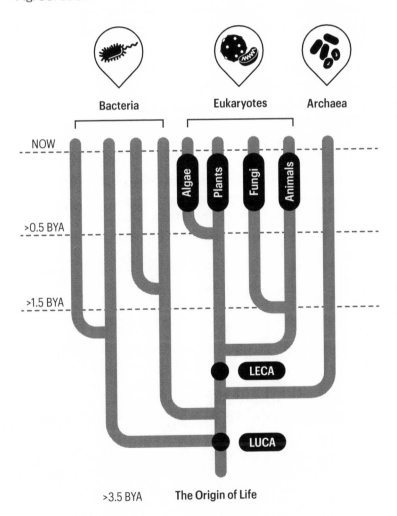

The implications are profound. Perhaps the greatest is the realization that all the cells in all our bodies are descended from LUCA. And although LUCA evolved about four billion years ago, most of the same "programs" that it developed are still right there in our cells, stored safely away in our DNA. Now it's true that LUCA's descendants continued to evolve and are much more sophisticated than LUCA was. But all the basic machinery was there in LUCA and is still with us today. This means that understanding LUCA, and her descendants, can help us better appreciate our own bodies and critical topics such as health and disease, aging and death. In the next chapter, we'll fast-forward almost two billion years to meet LUCA's most famous and most improbable descendant—LECA.

References and Further Reading

N Lane, JF Allen, and W Martin (2010). How did LUCA make a living? Chemiosmosis in the origin of life. *Bioessays*, 32(4): 271–80.

V Sojo, A Pomiankowski, and N Lane (2014). A bioenergetic basis for membrane divergence in archaea and bacteria. *PLOS Biology*, 12(8): e1001926.

V Sojo et al (2016). The origin of life in alkaline hydrothermal vents. *Astrobiology*, 16(2): 181–97.

MC Weiss et al (2016). The physiology and habitat of the last universal common ancestor. *Nature Microbiology*, article number 16116.

4

Life 2.0

"Thus a eukaryotic cell may be thought of as an empire
directed by a republic of sovereign chromosomes in the nucleus.
The chromosomes preside over the outlying cytoplasm in
which formerly independent but now subject and degenerate
prokaryotes carry out a variety of specialized service functions."

GUNTHER SIEGMUND STENT

IN THE last chapter, we learned how life might have evolved in undersea hydrothermal vents, ultimately figuring out how to enclose itself in a fatty bubble and sail off into the ocean. This resulted in one type of cell that was clearly superior to all its competitors: LUCA, the last common universal ancestor of all life on earth. LUCA quickly spread throughout the ocean, betraying its rocky origin by maintaining little iron-sulfur clusters at the core of increasingly complex enzymes used to speed up critical chemical reactions.

If we consider LUCA to be the first version of life, we can label the next major revision, archaea, as Life 1.1. As we discussed previously, archaea fall into the same family as bacteria—prokaryotes. Prokaryotes lack internal membranes like the ones later cells use to surround their DNA. They also lack the cellular "power plants" called mitochondria that generate ATP. Even through a microscope, it's difficult to tell the difference between bacteria and archaea. Examining more deeply, however, it's clear that archaea are a fairly significant upgrade over bacteria. They have tougher membranes that allow them to go places that bacteria simply can't—boiling geysers, frozen lakes, salty marshes, acidic ponds, and many others. They also have more modern-looking protein machinery for replicating their DNA.

After archaea hit the scene, innovation slowed down for a while. If we assume that Life 1.1 (archaea) was released about 3.5 billion years ago, it took another billion years for evolution to develop the next significant update—bacteria that could perform the magic of photosynthesis. This allowed them to utilize energy from the sun to turn their paddlewheels (turbines) instead of burning food molecules. They could roam even more widely because as long as they had access to water, carbon dioxide, and sunlight, they could survive. All three of these components were still present in abundance more than two billion years ago.

Life settled back into another semi-retirement for the next billion years. But like a rock band that toils in obscurity for a decade before stunning the world with a brand-new album, life came out with a major new release that left the critics speechless—Life 2.0.

Before we go into the specifics of this incredible new version of life, let's take a moment to gain some perspective. Bacteria and archaea, the headliners of the first iteration of life, are roughly the same size—about one micron in diameter. That's one millionth of a meter: so small that it takes a microscope to see them. They both have proteins embedded in their enclosing membranes that are able to pump protons out and then let them flow back in via the tiny turbine machine we've already discussed—ATP synthase. This turbine employs the force of the flow to squeeze a phosphate group onto ADP to make the high-energy ATP molecule used to power a variety of chemical reactions. Their DNA floats freely inside the cell with no enclosing membrane. Bacteria and archaea aren't big enough to swallow other prokaryotes, and their protein-studded membranes aren't designed for that task anyway.

Fig. 34: LUCA

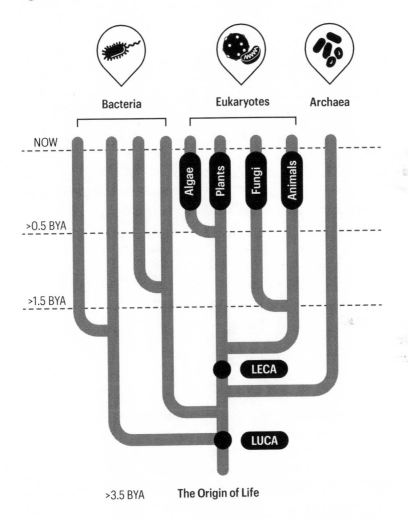

>3.5 BYA   **The Origin of Life**

By about 2.5 billion years ago, a few types of bacteria had evolved the machinery to use the energy from a photon of light to rip hydrogen atoms with their electrons off water. This is an impressive feat because, as we learned previously, oxygen is an electron miser. It takes a *lot* of force to wrest the two hydrogen atoms, including their electrons, away from the oxygen atom in a molecule of water. As these photosynthetic bacteria began to proliferate in the oceans, oxygen started to accumulate. For the first time in earth's history, detectable levels of molecular oxygen (O_2) began to build up in both the oceans and the atmosphere.

A Clever Way to Avoid Oxygen Poisoning

Many bacteria and archaea suffered mightily from the presence of this gas, which is life-giving for us but toxic for them. Some of their most critical reactions break down in the presence of oxygen. Undoubtedly, the death of huge numbers of prokaryotes resulted from their failure to adapt quickly enough to this newly oxygenated world. Of course, the bacteria that learned the trick of photosynthesis had to come up with ways to protect themselves from the by-product they were now generating. This led to the formation of enzymes like catalase that can take care of especially corrosive forms of oxygen, the so-called reactive oxygen species (ROS) or "free radicals" that you might have heard of.

Imagine that you're a bacterium or an archaeon that hasn't evolved the capacity to deal with oxygen. For you, oxygen is a deadly poison, as deadly as cyanide is for us. Just like cyanide, oxygen disrupts your metabolism and prevents you from generating energy. What do you do in a world where oxygen levels are going up just about everywhere? One strategy would be to retreat to the ocean depths or another place where oxygen is relatively scarce. But here's a radical alternative: Make friends with other bacteria that have evolved new methods of dealing with this lethal new gas, even when it's dissolved in water. It turns out that over the eons, some bacteria evolved a different kind of energy production scheme that could actually put hydrogen atoms onto oxygen and turn it into water! These bacteria could start with hydrogen-rich compounds like hydrogen sulfide (H_2S) that are analogous to the uphill water in the waterwheel from the last chapter. They learned how to pass these electrons down a kind of molecular bucket brigade.

Fig. 35: Molecular bucket brigade

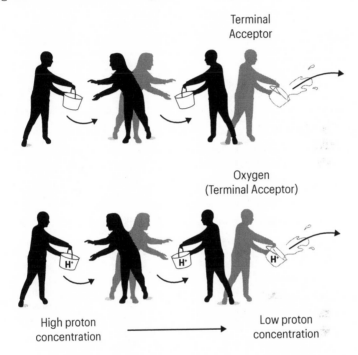

Terminal
Acceptor

Oxygen
(Terminal Acceptor)

High proton   Low proton
concentration   concentration

The "terminal acceptor" in figure 35 is molecular oxygen. When an oxygen atom accepts hydrogen atoms, including their electrons, we have the chemical reaction:

$$4H+ + 4e- + O_2 \rightarrow 2H_2O$$

What's hard to gather from the fire brigade analogy is that each time one person pours the contents of his bucket into the next person's bucket, a proton is pumped across the cell membrane in order to maintain the proton gradient we've discussed previously. As a single cell susceptible to oxygen poisoning, you love to hang out with bacteria that can perform this almost magical feat of converting oxygen into water. Imagine being near a nuclear reactor with a friend who could turn radiation into ice cream! It's not hard to see how oxygen-intolerant cells would love to develop close relationships with bacteria able to deal with oxygen in this way. This might well have led to the following sort of association:

Fig. 36: Origin of the eukaryotic cell

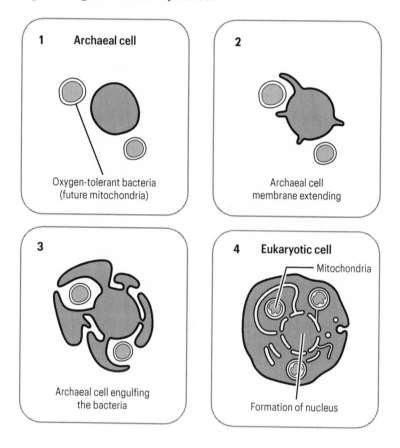

This is the model proposed by scientists David and Buzz Baum. It starts off with the oxygen-tolerant bacteria living close to a primitive archaeal cell unable to deal with oxygen. Over time, the archaeon extends its membrane to hold on to the bacteria and eventually move them permanently inside. This model might even explain how the nuclear membrane came to be.

The benefits of this arrangement to both parties are obvious. The bacteria get to live in the sheltered interior of the archaeon. The archaeon is happy to provide nutrients to its bacterial guests as long as they'll mop up any available oxygen and turn it into water. Over millions of years, their descendants make this a permanent arrangement, and the eukaryotic cell is born—Life 2.0.

To start with, the internalized bacteria are still capable of living on their own. They have their own DNA and make their own proteins. Over time, however, the bacteria die, and bits of their DNA are incorporated into the DNA of the host archaeon. It starts to make the proteins encoded by this DNA, and the bacteria evolve pores through which to absorb these proteins. The corresponding bacterial DNA gets less and less use until finally the internalized bacteria stop bothering to even maintain it. After all, if the proteins they need are being manufactured by their host, why keep making them? Today, after a couple of billion years of refinement, the descendants of these bacteria maintain the recipes for only about a dozen protein-coding genes out of the five hundred or so they started with. All the rest have been moved into the nuclear genome of the new eukaryotic cell. The final step was for the host archaeon to learn how to drill holes into its bacterial guests in order to steal some of their ATP. Today, the descendants of these original symbiotic bacteria still live in the cells of almost all eukaryotes and are known as mitochondria.

It's impossible to overstate the importance of this development. Prokaryotic cells are always scrounging for energy. Their membrane surfaces can hold only so many machines for pumping out protons or producing ATP. And as a cell gets bigger, its volume grows much faster than its surface area, so beyond a certain point, there's just not enough membrane surface to support further growth. This stark fact relegates prokaryotic cells—bacteria and archaea—to evolutionary dead ends. Although they've been quite successful spreading all over the earth, they just don't have enough energy to grow any bigger or more complex than they have been for billions of years. Fossilized remains of cyanobacteria from two billion years ago look identical to modern bacteria.

Enter the Last Eukaryotic Common Ancestor

Life 2.0, the merger of an archaeon with oxygen-tolerant bacteria, blew that surface area constraint away. The new eukaryotic cell no longer needed its outer membrane for energy production. That job could now be fulfilled by hundreds or thousands of mitochondria using their membranes to generate energy just like their bacterial forebears. Imagine an eighteenth-century wooden ship suddenly outfitted with a nuclear reactor. The wooden ship was limited in size and complexity because it had to derive all its energy from its sails. But with a nuclear reactor providing the power, even eighteenth-century engineers would have been able to make some amazing modifications.

And unlike such a ship, the eukaryotic cell had more than two billion years to upgrade its other components. Over the eons, mitochondria were given more responsibilities within the cell beyond mere energy generation. As semiautonomous outposts far from the nucleus, they became early warning systems. Today, our cells rely on mitochondria to detect various stressors and to signal the nucleus. Mitochondria are also executioners, able to trigger a built-in self-destruct mechanism when good cells go bad, as in cancer.

Just as we can trace all cellular life back to LUCA, the last universal common ancestor, we can reasonably infer that all eukaryotic life arose from LECA, the last eukaryotic common ancestor. Again, we arrive at this conclusion because of the striking similarities across all eukaryotic cells. It doesn't matter if we take them from flowers or fish, moths or monkeys, yeast or yaks. All eukaryotic cells are so similar that they must have derived from a common ancestor. In addition to having their DNA enclosed by a nuclear membrane, they all organize their DNA by winding it around protein spools called histones. They all either have mitochondria or had them but lost them along the evolutionary path. How can we tell that a eukaryotic organism used to have mitochondria? Because the story is right there in its DNA. We can see the mitochondrial genes that were transferred to the nucleus and conclude that a few rare single-celled eukaryotes got rid of their mitochondrial guests at some point in the distant past, probably because they didn't need them due to unusually low energy requirements.

Fig. 37: LECA/eukaryotes

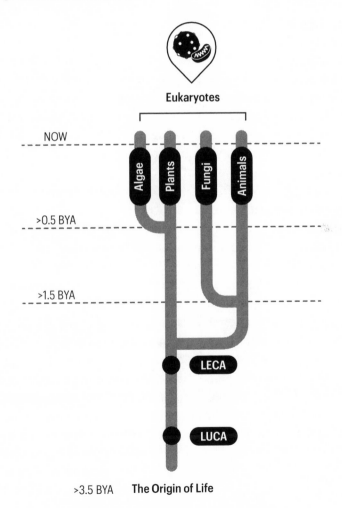

>3.5 BYA   **The Origin of Life**

There are many more similarities across different types of eukaryotic cells than we can possibly cover here. So, suffice to say that the evidence is overwhelming that the sort of symbiosis between an archaeon and oxygen-tolerant bacteria we described earlier took place only once on our planet. That makes such a union an extremely unlikely event and a big barrier for life to cross. If and when we find life on other worlds, it's most likely to be prokaryotic in nature.

The Hunters Take Over

The development of the eukaryotic cell led to an exuberant explosion of life on our planet. Freed from energy limitations, eukaryotic cells grew much larger than their prokaryotic ancestors. Eukaryotic cells are roughly ten times the diameter of prokaryotic cells, and their volumes can be thousands of times greater. And with their outer membranes no longer devoted to supporting the maximum possible number of ATP synthase machines, eukaryotes were free to experiment with using these membranes for other purposes. In particular, they evolved to become hunters. While it's true that prokaryotes take in high-energy molecules from dead and dying kindred, they aren't able to "eat" other organisms in the way eukaryotes are; they are too small to swallow other bacteria and archaea. Also, their membranes are full of protein machines needed for energy production and thus don't have free stretches that might wrap around other organisms.

Eukaryotes, on the other hand, are carnivores in the sense that they can physically engulf and digest other organisms—both prokaryote and eukaryote. Their large size makes it relatively easy for them to swallow smaller cells, and their membranes are much more flexible because they don't need to put all their energy-production machinery there. As we'll see later, that job is taken over by the mitochondria that were previously free-living bacteria. The process of one cell "eating" another is called *phagocytosis* and is a hallmark of eukaryotic cells. The type of white blood cell called a macrophage is an excellent example of a phagocytic cell that our immune system uses to literally consume invading bacteria and even our own dying cells.

I'm sure all this sounds like a total win for LECA and her descendants. They gain an incredible new power source that allows them to grow thousands of times larger, hunt and consume other cellular prey, develop more complex genomes, and generally become bigger and badder than any life the earth had ever seen. But as fiction has proved to us many times, there's always a downside to selling your soul for infinite power. LECA didn't know it, but she was striking a Faustian bargain that would plague her eukaryotic clan for the rest of time.

References and Further Reading

D A Baum and B Baum (2014). An inside-out origin for the eukaryotic cell. *BMC Biology*, 12: 76.

L Eme et al (2017). Archaea and the origin of eukaryotes. *Nature Reviews Microbiology*, 15: 711–23.

N Lane (2015). *The Vital Question: Energy, Evolution, and the Origins of Complex Life*. WW Norton and Company, New York.

5

Life 3.0

"A band of bacterial brothers, swigging ATP
with some others. In a jocular fit, they laughed 'til
they split. Now they're all microbial mothers."

RICHARD COWEN

A S WE'VE seen in previous chapters, the first version of life came in the form of simple, single-celled prokaryotes— bacteria and their tougher cousins, the archaea. They collectively ruled the planet for well over a billion years until life came out with its next major release. Life 2.0 resulted from the symbiosis between an archaeon and some oxygen-tolerant bacteria that it allowed to set up shop inside its membrane so they would mop up the growing environmental pollutant, molecular oxygen (O_2). This created a new type of cell—the eukaryote— with thousands of times more energy at its disposal. Suddenly eukaryotes were free to grow huge in comparison to their prokaryotic predecessors. And they developed the ability to hunt, swallow, and digest other cells, turning them into the world's first carnivores. As impressive as this was, life had yet another trick up its sleeve.

For the first billion or so years of their existence, eukaryotes continued to live as individuals, much like gunslingers in the Old West. They trusted no one, and no one trusted them. For a billion years, they concentrated on letting evolution tinker with their DNA to make them even better. If the first eukaryotic cell, LECA, was like a wooden ship with a nuclear reactor bolted on, then the eukaryotes at the end of a billion years of evolution were like the starship *Enterprise*. They preferred proton power to dilithium crystals, but both were incredibly sophisticated. As an aside, it's funny to think about how life originated in mineral crystals, and even science fiction continues that theme.

Beginning about a billion years ago, there are indications that eukaryotic cells were starting to form complex federations. Sometimes this was just different types of eukaryotes living in the same region as a loosely allied colony. This allowed some free trade that benefited everyone. For example, one type of eukaryote might produce compound A as waste, while another eukaryote might use A as a food source. This would obviously encourage the second type of eukaryote to hang around the first. This association would be strengthened even further if the first eukaryote used compound B as a food source and if the second eukaryote produced this same compound as waste. Now we have a pleasantly codependent relationship—one makes what the other needs and vice versa. Although prokaryotes evolved similar sorts of codependencies, they never took the concept any further. But eukaryotes kept going. Human trade started off with isolated clans exchanging goods via simple barter but evolved into complex nation states and trading exchanges of dizzying complexity. Eukaryotes followed a similar social trajectory.

Perhaps it was their greater energy supply. Or maybe it was the way mitochondria freed up their membranes for pumps, transporters, and adhesion molecules. Whatever the reason, there are signs that by about one billion years ago, eukaryotes were making the leap to multicellularity. This culminated in a frenzy of multicellular experimentation that began about a half-billion years ago, a period called the Cambrian Explosion. We see for the first time in the fossil record all sorts of complex creatures, many of which had begun to cover themselves in shells for protection.

New Body Plans

The first multicellular animals were little more than bags with one opening that served as both mouth and anus. This allowed them to scoop up large volumes of seawater or soil to extract the nutrients and then eject the waste out the same orifice. These early animal models had two distinct features—radial symmetry and a two-layer organization. The jellyfish is the classic example of radial symmetry. Just like with a dome, any vertical slice that goes through the exact top point will look the same as all other such slices. Two-layer organization just means that if you cut into the creature and look under a microscope, you will see two different types of cells—outside and inside. Think of the outside type as skin and the inside type as stomach. Both were specialized for different purposes.

By five hundred million years ago, new body plans had come on the scene. The most successful of these involved bilateral symmetry and a three-layer architecture. Almost all the animals you can think of are bilaterally symmetric—that is, their two halves are mirror images of each other. Just think of a human body with its left and right sides, each with an eye, an ear, an arm, a leg, and so on. The three-layer cellular architecture in humans consists of the outer ectoderm (skin), the internal endoderm (lining of the gut), and a layer of mesoderm (muscle, bone, blood vessels, etc.) between them. As it turns out, the ectoderm is also the origin of the brain and spinal cord, while the endoderm gives rise to internal organs such as the lungs and pancreas.

It's important to step back and contemplate the many engineering problems organisms faced as they made the move to multicellularity. Every cell in such an organism was a descendant of the free-living LECA, used to living on its own and fending for itself. No wonder it took hundreds of millions of years to evolve all the features required to get millions—or trillions—of previously independent cells to live peacefully as part of a complex collective. And you just know that not all cells peacefully handed over control to the central government. There had to have been rogues and rebels who decided to exploit the situation for their personal advantage. In fact, as we'll see later, this is still a problem lo these hundreds of millions of years after the initial mergers.

Fig. 38: Germ layers

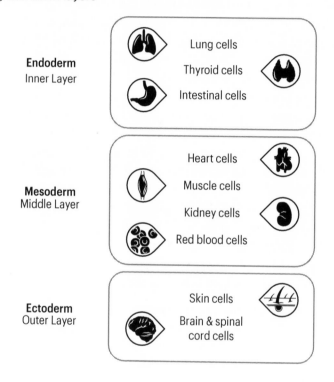

Germline and Soma

The multicellular creatures that began to appear about a half-billion years ago made an interesting compromise right off the bat. Obviously in single-celled organisms every cell has the potential to divide and pass on its DNA to the next generation. Multicellular organisms decided to specialize into two different cell lines right from the beginning. A few cells would become the seeds of the next generation. These so-called *germ cells* are what we know as eggs and sperm. Early in its development, each animal takes a few cells and puts them through a few rounds of a special kind of cell division. For most cells, division involves copying all the DNA and then splitting into two cells each with a full set of the DNA recipe books (chromosomes). The result is two cells that are more or less identical copies of the original.

Fig. 39: Mitosis

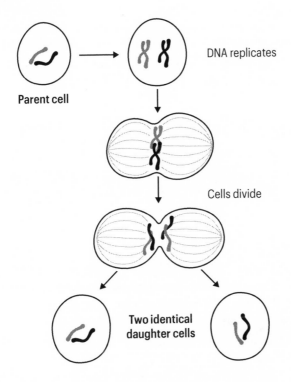

Parent cell

DNA replicates

Cells divide

Two identical daughter cells

For germ cells, the division process is a bit different. After replicating its DNA, the germ cell actually splits *twice*. The result is now four cells, each with only one copy of each chromosome instead of the usual two. When a sperm fertilizes an egg, the result is once again a cell with two copies of each chromosome, albeit in a new, unique combination.

The non-germ cells that account for far more than 99 percent of the total are called *somatic cells* and constitute the "soma" or body. Think about this bargain from their standpoint. They have to give up all hope of immortality. At best, they'll be able to divide a few times and contribute to the development of the organism, but they have no hope of passing on to the next generation. One explanation of aging is called the disposable soma theory. The idea is that organisms always have to make hard choices because of limited energy, and every bit of energy invested in the soma (body) is energy not applied to the germline. In a sense, the body is a disposable vehicle for the germ cells; its only purpose is to find a mate and pass its germ cells down to posterity. After reproduction, the body no longer really matters, so less effort is put into maintaining it. Ultimately, we age and die, but our DNA lives on in our offspring.

Although there are some problems with this theory, it does emphasize that our bodies are secondary in importance to our eggs and sperm, at least from evolution's standpoint. Our somatic cells do all of the day-to-day work. They have to actively use their DNA to make new proteins, which means constantly opening up their DNA recipe books and exposing them to many potential dangers: ultraviolet radiation, reactive oxygen species, chemicals from pollutants, etc. In contrast, the germ cells sit in a sort of suspended animation that offers far more protection from environmental hazards. Their DNA stays nicely coiled around their histone spools, and they rely on old-fashioned glycolysis to produce energy. This protects them from the reactive oxygen species spewed out by mitochondria during aerobic respiration. But more on that later...

Adhesion Molecules and Growth Factors

While unicellular creatures can move freely and independently, multicellular animals must tightly control every cell's mobility. The first step was for cells to develop adhesion molecules—kind of like Velcro made of proteins or carbohydrates. Cells insert many such molecules into their outer membranes. These adhesion molecules not only help the cell keep its place within the body, they also provide information from the cell's local environment. Movement is so strictly controlled within multicellular creatures that any cell becoming completely detached from its neighbors will soon auto-destruct and die. The license to travel within the federation is limited and regulated. A few white and red cells make circuits through the blood or lymphatic systems but eventually snuggle up to cells in lymph nodes, bone marrow, or other sites to regain contact with other cells.

Similarly, unicellular creatures make their own personal decisions about when to divide. Obviously that approach would be disastrous in a multicellular organism, so each cell must listen for commands to know when it's time. These commands come in the form of chemical messages called growth factors. When a sufficient quantity of growth factors interacts with receptors on the cell surface, the cell activates its division program to copy its contents and split in two. With trillions of cells, it's easy to see how this process could go wrong. We see the results in cancer—cells that decide to secede from the union to form their own colony.

Stem Cells and Tissues

The development of a complete animal from a fertilized egg is a breathtakingly beautiful process. Once fertilized, the egg enters a phase of frenetic activity. The single cell that is the fertilized egg first divides into two identical cells. Those then divide into four, and those four into eight. After a few rounds of such divisions, the cells form a ball called a *blastocyst*. A special group of cells on the inside of the ball (the inner cell mass) act as *stem cells* that will generate the rest of the body.

Fig. 40: Stem cell

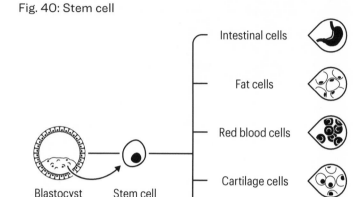

In some ways, stem cells are similar to the germ cells we described earlier. We can think of stem cells as germ cells for the tissues of the animal under construction. During the infancy of the animal, they must divide frequently to generate the trillions of cells that will make up the adult. After that, like germ cells, they go into a kind of dormancy during which they sip energy and do all they can to protect their DNA. One of the special attributes of stem cells is that they can divide in two different ways. One stem cell can divide into two identical stem cells to grow the stem cell pool. Or it can divide asymmetrically, with one daughter staying a stem cell and the other becoming more specialized down a particular lineage (e.g., liver, heart, brain, etc.).

Fig. 41: Stem cell dividing

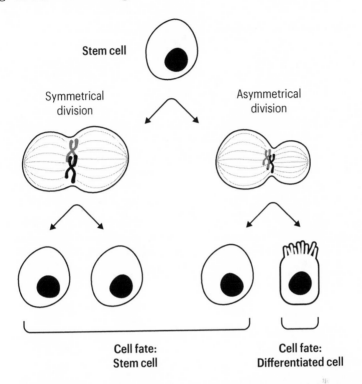

Most tissues in animal bodies have multiple levels of special-
ization. For example, the cells lining the gastrointestinal tract live
only about a week and then must be replaced. Stem cells don't
directly generate the fully differentiated enterocytes (cells that
line the intestines). Instead, a stem cell will produce what's called
a progenitor (or "transit amplifying") cell that is more specialized
than its parent but still retains some "stemness." This cell can
divide to form slightly more specialized cells and so on. Eventually we reach the terminally differentiated cells that form the
bulk of each organ.

Soon we'll be talking about aging. What do you think could go
wrong in this architecture as we get older?

Fig. 42: Stem cell to mature cell

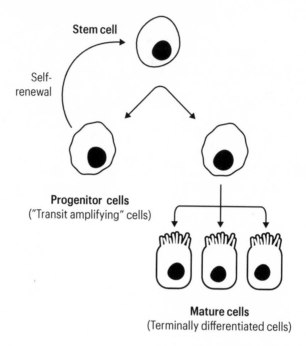

Stem cell

Self-renewal

Progenitor cells
("Transit amplifying" cells)

Mature cells
(Terminally differentiated cells)

The Circulatory System

The first multicellular organisms were small enough that they could rely on simple diffusion to get oxygen to all their cells and eliminate waste products from them. As these organisms grew larger, however, diffusion wasn't enough. The organisms slowly started to develop a circulatory system that could deliver oxygen and nutrients while carrying away carbon dioxide and other waste products. As figure 43 illustrates, the circulatory system became more and more sophisticated over the course of evolution. Mammals, including humans, have essentially two linked circulatory systems that meet in a four-chamber heart. One system takes oxygen-poor blood from one part of the heart (right ventricle) and pushes it through the capillary system in the lungs in order to bring in fresh oxygen and give dissolved wastes like carbon dioxide a chance to escape. The second system takes oxygen-rich blood from the lungs, pushes it out to the body, and then collects oxygen-depleted blood from around the body and returns it to the heart (the right atrium) to make another circuit.

Fig. 43: Pulmonary diagrams x4

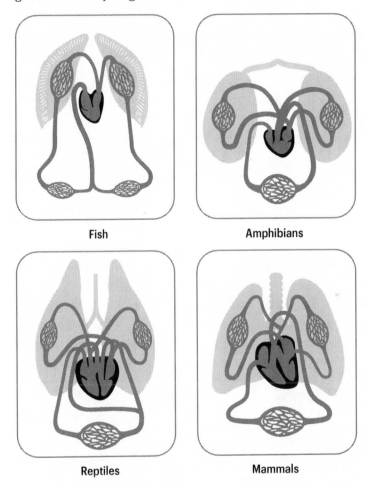

Fish

Amphibians

Reptiles

Mammals

It's fascinating to view the progression from fish to amphibians to reptiles to mammals. These increasingly complicated circulatory designs provided the oxygen and nutrients that allowed animals to get much larger and develop more and more complex behaviors.

As with any complex feature, there are downsides to such a circulatory system. Imagine the force that *endothelial cells* (cells that line the inside of blood vessels) are subjected to as a torrent of blood is pushed by them every second. The blood contains a variety of cells, proteins, and other molecules that bump violently against the endothelial cells. Over the years, these shear forces take their toll. Both the endothelial cells and the smooth muscle and connective tissue cells that surround them become less resilient. This is one reason that high blood pressure is so harmful if not treated.

The Nervous System

As animals grew larger, another problem they faced was communication. Whether it's Rome or an orangutan, a federation needs to be able to stay in touch with its farthest reaches. Small organisms were able to rely on simple chemical messengers that could travel from cell to cell like the old pony express. With larger, more complex body plans, a new approach was needed. The nervous system was a marvelous invention that turned to a much faster form of transmission, using a special type of cells—*neurons*—that evolved to send information over long distances with electricity.

Charged particles can travel down a conductor much faster than a chemical can diffuse through a watery medium. As we've seen, even early cells learned how to create and maintain proton gradients by actively pumping protons out of the cell. Neurons adapted the pumps and channels involved in this process to another purpose—communicating a signal from one end of a long cell to another. The signal could then jump the gap between cells using slow chemical signaling and keep right on going. This approach has evolved to the point that in a large mammal like a giraffe, the pain signal from hurting a foot can travel several meters to the brain in a fraction of a second.

Early nervous systems were quite primitive. Their main purpose was to implement reflex loops. Come into contact with something hot, and an automatic defensive reaction causes the animal to move back. There's no thinking involved in these sorts of reflexive actions, so no central brain is required. As animal body plans became more complicated, however, a central processor was needed. Evolution almost always involves layering new things on top of an existing infrastructure, so reflex loops still exist today. Step on a nail, and the pain will cause you to withdraw your foot before you even realize what happened. However, building connections between all parts of the body and a central brain offers unlimited opportunities for more complex behaviors.

Before the brain, evolution had to program all behaviors into the genes. We can think of the development of the complex nervous system as moving behavior up one level. It's similar to the difference between hard-coded computer chips and software-based systems. The hard-coded chips can only do a few things that have been anticipated in advance. Software running on a programmable processor, however, can be updated automatically thereby allowing new behaviors to be developed and deployed independently of the hardware. With the appearance of a brain and nervous system in animals, evolution reached a new level. Now organisms could "evolve" during their lifetimes and be able to deal with a much greater variety of dangers and opportunities than simpler hardwired creatures.

The only real downside to a brain and nervous system is all the energy it devours. Neurons are extremely energy-hungry, and complex clusters of neurons like the brain are even more so. Even though it accounts for only about 2 percent of the mass in the human body, the brain consumes more than 20 percent of its total energy. The disposable soma theory mentioned earlier would predict that this exorbitant investment might be dialed back after the reproductive period of an animal's life, and this is what we see in practice.

Energy Production

A critical compromise had to be negotiated when cells cooperated to become bodies. Until this time, every cell had been responsible for procuring its own food and generating its own energy. This had to change in a multicellular federation. Only some cells would be in contact with the external environment in a way that would allow them to absorb high-energy molecules like glucose that could be consumed as food. All the other cells would have to wait to be supplied. Very early on in the development of animals, an architecture evolved to address this challenge. Some cells would absorb food molecules and pump them into the bloodstream. Other cells would monitor the bloodstream for the presence of glucose and send a signal to the rest of the cells when this precious molecule was available. This signal would cause those cells to open up a protein pore in their membranes that could grab the glucose from the blood and pull it inside the cell.

In modern animals, this arrangement has evolved to incorporate an impressive division of labor. Specialized cells that line the stomach and intestines break down large chains (fats, proteins, and carbohydrates), absorb the individual components (fatty acids, amino acids, and sugars), and pump them into the bloodstream. The beta cells of the pancreas act as sensors, monitoring the blood for the presence of glucose. When they detect it, they secrete insulin into the blood. The insulin circulates in the body to alert cells in muscle, brain, kidney, and other tissues to the availability of this high-energy molecule.

The liver also plays a critical role in metabolism. It, too, monitors the bloodstream for insulin and glucose. When those are plentiful, it stores glucose away in long branching chains called glycogen. This is the first reserve that will be tapped the next time energy stores begin to drop. In this way, the liver helps maintain a fairly constant level of glucose in the blood—absorbing excess glucose when there's too much and breaking down glycogen to pump glucose into the blood when levels are too low. Muscles are also able to store excess glucose as glycogen and to dip into these glycogen stores when a quick burst of activity requires it.

Of course, mammals have a second form of energy storage. The liver can also use excess glucose to synthesize fatty acids. These long carbohydrate chains are packaged three at a time with one molecule of glycerol to create triglycerides—the storage form of fat. These triglycerides are loaded onto protein carriers and pushed into the blood where they can be absorbed by fat (adipose) cells.

Muscle cells are among the most avid energy consumers in the body, tapping into food molecules to generate the required energy. First, they use the insulin signal to take available glucose molecules from the blood. If there's not much glucose at a time when they're being asked to contract, muscle cells will tap into their own glycogen reserves and use the glucose there. When that runs dry, the liver will signal to fat cells around the body that they need to start breaking down their triglycerides and secreting fatty acids into the blood. Cells around the body will absorb these fatty acids and transport them into their mitochondria to be burned as fuel. So the general order of utilization of fuel molecules is (1) recently consumed food, (2) glycogen, (3) fat.

Note that this third fuel source doesn't pass efficiently through the barrier surrounding the brain (the blood-brain barrier), so the liver absorbs some of these fatty acids from the blood and chops them into smaller pieces called ketones, which can easily penetrate the blood-brain barrier and become an important fuel source for the brain during fasting periods. Both intermittent fasting and ketogenic diets are ways to force the body into a fat-burning mode, and both lead to an increase in blood ketones. We'll talk more about these dietary techniques later.

References and Further Reading

JT Bonner (1998). The origins of multicellularity. *Integrative Biology*, 1(1): 27–36.

E Libby and WC Ratcliff (2014). Ratcheting the evolution of multicellularity. *Science*, 346(6208): 426–27.

PART TWO

Aging and Death in Complex Organisms

6

The Dark Side of Progress

ITTLE COULD LECA have known the train of events she was about to set in motion. She had evolved from the union between an archaeon and an oxygen-tolerant bacterium. The result was a new type of cell with almost unlimited power and immense potential. She had been given the choice between staying an archaeon forever or transforming into something completely different and largely unknown. As an archaeon, she was essentially immortal. She could make copies of herself at any time to create an unending lineage. But the archaeon made a deal with the devil, resulting in a new type of turbocharged cell with essentially unlimited power. LECA was the result of this Faustian bargain. While eukaryotes were a new form of life with dramatically greater energy capacity, the unholy union required to produce them was set to unleash a wave of death, disease, and destruction, the likes of which the earth had never seen.

Let's consider the fundamental problems resulting from two genetically different organisms trying to live as one. Imagine having a computer run two different operating systems at the same time on the same processor. This is possible today with advances in context switching, virtual machines, and other safeguards. But it would have been unthinkable in the early days of the PC industry and would have resulted in a computational disaster. The early eukaryotes must have been similar messes. Each such cell had a nucleus with DNA derived from the host archaeon. This DNA encoded not only the recipes for the various proteins that ancestral cells needed to make, but also the various "scripts" or "programs" that were executed under different conditions. For example, one program might specify the production of a set of proteins in a defined sequence to deal with a sudden increase in environmental temperature (so-called heat shock). Another program might kick in when the cell found itself in acidic locations. The various programs the DNA of the host archaeon encoded had evolved over hundreds of millions of years with one overarching goal—keep the archaeon alive and help it reproduce.

The same thing was true of the DNA in the bacterial guest. These bacteria seem to be closely related to a modern group of bacteria called alphaproteobacteria. As free-living beings, they had their own agenda, which prioritized their survival and reproduction. Each of these entities—the host archaeon and the guest alphaproteobacteria—had evolved to look out for its own interests. Put one or more bacteria inside an archaeon, and what do you expect to happen? The guest bacteria will certainly try to reproduce, potentially even at the expense of the host cell. The archaeon will continue to look out for its own interest, even to the detriment of its bacterial guests. It's no wonder that this union was successful only a single time in the history of life on this planet. In fact, it's a miracle that it happened at all.

We can surmise that the original alphaproteobacterial symbiont was at least oxygen-tolerant and probably used oxygen as the final electron acceptor just as most modern eukaryotic cells do, producing water as a by-product. It's time to dig into that process to understand how eukaryotic cells rely on the descendants of these bacterial symbionts, now called mitochondria, to make the most of the cell's energy.

How Mitochondria Produce Energy

Let's start by understanding how mitochondria are organized. A modern eukaryotic cell can contain anywhere from a handful to thousands of mitochondria. Muscle cells, for example, are packed full of mitochondria due to the large amounts of energy these cells need to operate. Other tissues with high energy requirements include those of the liver, kidney, and brain.

A mitochondrion is an oval structure surrounded by two membranes. The inner membrane corresponds to the cell membrane of its bacterial ancestor. The outer membrane comes from the cell membrane of the archaeon host cell, as illustrated in figure 44.

Fig. 44: Mitochondria origin

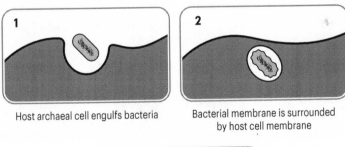

Host archaeal cell engulfs bacteria   Bacterial membrane is surrounded
                                                        by host cell membrane

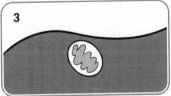

Bacterial membrane becomes **inner mitochondrial membrane**
Host cell membrane becomes **outer mitochondrial membrane**

Over time, the inner membrane began to develop many folds to increase its surface area and allow for the insertion of more protein complexes for energy production. The portion of the mitochondrion inside the inner membrane is called the matrix.

Fig. 45: Mitochondria anatomy

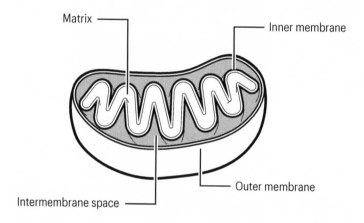

But exactly how do mitochondria produce energy? Although the details are complicated, the overall approach is elegantly simple. It comes down to this: Mitochondria have protein enzymes that pluck hydrogen atoms and their lone electrons off carbon-based compounds rich in these chemical resources. They then pass the electrons downhill. Think back to our waterwheel example and consider a single droplet of water. First, it hits the top blade in the waterwheel, the one closest to the uphill water source. This force (along with that of many droplets) pushes the blade and causes the wheel's shaft to turn. It's this force that is used to grind wheat or generate electricity.

As the waterwheel turns, that first droplet falls and hits the second blade. The water droplet has less potential energy now because it isn't as high as it originally was, but it still transmits force to the second blade and causes the wheel to turn some more. Depending upon the size and design of the waterwheel, a given droplet of water may strike several blades before it eventually falls to the ground, its potential energy exhausted and transformed into the kinetic energy turning the wheel.

Fig. 46: Water wheel kinetic energy

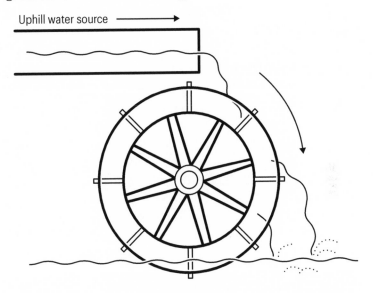

It's important to see the two separate but related things happening here. First, we have the potential energy in the uphill water droplets being converted into kinetic energy to turn the waterwheel. Second, the rotation of the waterwheel turns a stone to grind wheat. Ultimately, energy is converted from one form (potential) to another (kinetic). The way that mitochondria generate energy is amazingly similar. The mitochondria start with a chain of carbon atoms with some hydrogen atoms bonded to some of the carbons. The classic example is pyruvate, shown in figure 47.

Fig. 47: Pyruvate

Pyruvate

As you can see, pyruvate is a string of three carbon atoms. The right end as shown in the diagram is essentially a molecule of carbon dioxide. It's the left end that contains the energy—three hydrogen atoms, each sharing an electron with the leftmost carbon. By now, when you see that pattern, you should start feeling hungry and your brain should be saying "food!"

Another way to think of pyruvate is as half a glucose molecule. Glucose is a six-carbon chain. Let's not worry about where the pyruvate comes from right now, but just think of it as an electron-rich energy source for mitochondria. But how to get at that energy?

Well, the simple way would be to burn it. This is what happens when we burn wood in a fire. Oxygen molecules in the air start ripping electrons away from the hydrocarbons in the wood. This is a violent reaction that releases tons of heat. It also serves as a kind of chain reaction because burning one part of the wood makes adjacent parts more likely to react with oxygen, thus spreading the reaction. Although cells (and mitochondria) like to stay warm, they disintegrate if they get too hot, so simply burning pyruvate won't do the job for us. Cells had to come up with a cleverer method of energy access.

Let's go back to our waterwheel. What if we took the wheel out of the picture and just let the uphill water fall to the ground? What would happen? Well, the potential energy of the uphill water would be converted quickly to kinetic energy. All that energy would be turned into heat when the water hit the ground. We could try to harness some of that heat to do something useful, but it would be hard. Most of the heat spreads quickly into the environment and is lost. That's why the waterwheel is so useful. It allows us to *gradually* siphon off potential energy as the water falls. The transit to each blade captures a bit more of the water's potential energy and allows the shaft to turn in a controlled manner. How can we do something similar with the chemical potential energy stored in pyruvate?

In essence, nature has evolved a molecular bucket brigade for electrons. As a hydrogen atom with its electron is plucked from pyruvate, it "falls" onto another molecule that is specialized in carrying electrons. Most commonly, this molecule is a coenzyme called nicotinamide adenine dinucleotide, NAD+ for short. NAD+ is very fond of electrons. In fact, it can accept a whole hydrogen atom with not only its electron but also another one stolen from pyruvate (both electrons in one of the C—H bonds) to become NADH. But the important point is that NADH is just fine with handing those electrons (and the lone proton) off to another molecule. So NADH acts as one of the cell's major *electron carriers*. Very little of the original energy in the C—H bond is lost when the H with two electrons is handed over to NAD+. We can think of NAD+ as the flume in the waterwheel in figure 46. A good flume is one that doesn't leak the water it carries to the wheel.

The "wheel" or "bucket brigade" in a mitochondrion is a set of four little protein machines collectively known as the electron transport chain (ETC). They have been numbered complex I through complex IV. You can think of each one as a blade on the waterwheel with complex I at the top, complex II a little lower, complex III lower still, and complex IV at the bottom. NADH passes off the two electrons and the proton that it stole from pyruvate to again become NAD+. Complex I receives both the electrons and the proton from NADH. As part of complex I, the electrons lose some of their potential energy just like water falling from one blade to another. Just as the waterwheel uses this energy to help turn the shaft, the proteins in complex I use the energy lost by the electrons to push the proton out of the mitochondrion's interior into the area between the inner and outer mitochondrial membranes—the intermembrane space.

Fig. 48: Electron transport chain

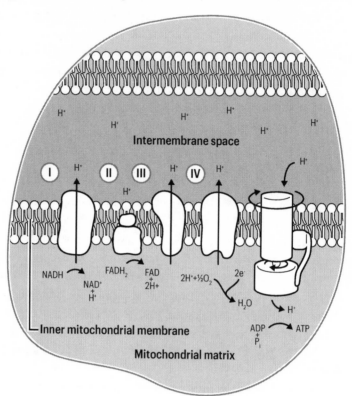

As you can see in figure 48, this process causes the intermembrane space to fill up with protons (H+) as electrons are passed from one "blade" to another. But what happens at the last stop? Just as the blades on the waterwheel have to continue back up once they've hit the bottom, we need to get the electrons out of complex IV so we can continue the bucket brigade. Otherwise, the whole thing would grind to a halt. Can you guess what the final "electron acceptor" is? What kind of atom do you know that is hungriest for electrons? That's right—oxygen.

Molecular oxygen (O_2) is waiting greedily at the end of the electron transport chain to accept electrons from complex IV.

When those electrons hop onto oxygen, that molecule becomes negatively charged and starts looking around for protons. As we'll see in a minute, even though we've been pumping protons out of the matrix, they eventually flow back in and are available to jump onto the negatively charged oxygen. So we end up with this final reaction:

$$2H+ + \frac{1}{2}O_2 \rightarrow H_2O \text{ (or } 4H+ + O_2 \rightarrow 2H2O)$$

The protons and electrons that are stolen from pyruvate eventually combine with oxygen to form water. The carbon atoms left over from breaking down pyruvate also combine with oxygen to form carbon dioxide, which we breathe out into the air. If this sort of reaction took place in the bacterial precursor to mitochondria, it's easy to understand why it would have made an attractive partner to an oxygen-phobic archaeon—the bacteria could turn toxic oxygen gas into harmless water!

ATP Synthase, an Amazing Energy-Producing Machine

Not to sound like a late-night infomercial, but wait, there's more! The mitochondrion doesn't just pump protons into the intermembrane space for nothing. The protons represent a new form of potential energy because of the strong difference in concentration this process creates between the intermembrane space and the matrix. Such a difference in potential energy can always be tapped if you're clever enough, and cells are endlessly clever. The ribosome that manufactures proteins according to genetic recipes is probably the most amazing machine in the universe, but ATP synthase (the machine on the far right of figure 48, after complex IV) is a close second.

Fig. 49: F_0 F_1 inner mitochondrial membrane

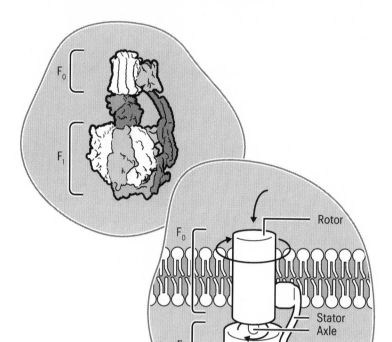

The illustration on the left in figure 49 shows the various proteins involved in this incredible machine, while the one on the right puts it into its context within the inner mitochondrial membrane. As electrons flow through the outer part (F_O) of the machine, the axle inside the mitochondrial matrix turns. It's hard to tell from the diagram, but the F_1 module has three chambers: One chamber contains a molecule of ADP, and one holds a molecule of phosphate. The force of the rotor turning squeezes these two together to form a molecule of ATP, which exits from the third chamber, and the whole process starts all over. It takes approximately three protons passing through the machine to generate one ATP.

The inner membranes of mitochondria are studded with hundreds of copies of complexes I through IV of the electron transport chain as well as ATP synthase. This is how the billions of mitochondria in the human body are able to produce roughly an adult human's weight in ATP every single day, which gets turned back into ADP as that body uses energy. We won't go into much more biochemistry here, but just understand that ATP is another form of chemical potential energy. It's like a bowling ball that we keep up on a chair. When we need to use its energy, we let it roll off and once again turn that potential energy into a form that can do work—like adding amino acids onto the end of a growing protein chain or causing muscles in the heart to beat.

So, what's the downside? So far, this seems like a fairy-tale marriage. The bride and groom mesh perfectly and live happily ever after. Unfortunately, nothing in life (or biology) is ever quite that simple. Considering the difficulties of this unlikely union, it's hard to believe that it didn't end in a bitter divorce long ago.

When the archaeon host first welcomed its bacterial guest, the two were independent organisms. Each had its own DNA blueprint specifying which proteins to make and implicitly how to use them. Even more importantly, each organism had its own agenda—to replicate. Imagine the poor archaeal host. It lets one alphaproteobacterium inside where it's nice, safe, and cozy. The interior of the archaeon is filled with food molecules and other building blocks—sugars, fats, amino acids, etc. So the bacterium does what it has evolved to do—grow and divide. Soon there are four alphaproteobacteria, then eight, then sixteen, and so on. If the alphaproteobacteria continue to divide, they'll eventually cause the host cell to burst, and that will be the end of the evolutionary experiment.

Perhaps the archaeon figured out a way to tell its bacterial guests to stop replicating. Or perhaps the alphaproteobacteria had some genetic mutation that served to limit their rate of replication. Regardless, we clearly have a conflict of interest that must be sorted out. In modern cells, mitochondria are signaled by the nucleus to divide, so they are under central control. That could not have been the case initially. Over time, this problem was mitigated by the transfer of protein-coding genes from the bacteria to the host DNA. Even so, the cores of the protein machines in the electron transport chain are still coded by the mitochondrial genome a couple of billion years later. This creates a coordination problem. When it's time for mitochondria to grow or to add more electron transport chain complexes, the following must happen:

1   The nucleus has to transcribe the hundreds of genes that code for most mitochondrial proteins.

2   The mitochondrion has to make the dozen or so proteins whose recipes are still stored in mitochondrial DNA.

3   The nuclear-coded proteins and mitochondrial-coded proteins must be combined inside the mitochondria in order to assemble the new complexes needed.

One problem with this approach is that the nucleus can't target its products to one specific mitochondrion. Instead, it simply increases its production and sends the results out to *all* the mitochondria in the cell. This makes it very difficult to match supply and demand exactly. A few mitochondria could soak up all the nuclear output and leave the rest under-supplied.

That tension between the needs of the cell and those of the mitochondria persists in our bodies to this day. For instance, it's thought that a large percentage of miscarriages are due to a mismatch between the mitochondria (which come from the mother's egg) and the DNA of the early fetus (which comes from both the mother and the father). If these aren't compatible, the body of the pregnant woman will usually detect the problem and abort the fetus. We can think of this as nature's screening process— try a different egg each month, use some of the father's sperm to fertilize it (combine DNA from the father and the mother), and see if the resulting fetus develops normally. If not, miscarry it and start over.

A Dangerous By-Product: Free Radicals
. . . .

Another downside to the archaeon-bacterium union lies with the latter's flirtation with oxygen. When the advent of photosynthesis led to an increase in molecular oxygen on earth, most organisms dealt with the problem by staying in deep ocean waters or other places where there was little free oxygen (not bonded to some other element). The newly formed eukaryote was now able to tolerate more oxygen-rich environments because its mitochondria—descendants of those early bacterial guests—could turn deadly molecular oxygen into harmless water.

But there's a catch. Oxygen is such an electron thief that it doesn't always wait politely in line at the end of the electron transport chain where it can be safely packed with enough electrons to construct water molecules. Some oxygen molecules jump the line and steal electrons from one of the first three complexes. The result is not water but a dangerously unstable "free radical." These reactive forms of oxygen come in three flavors—superoxide, hydrogen peroxide, and hydroxyl radical. Hydroxyl radicals are formed when hydrogen peroxide reacts with iron in a process called the Fenton reaction, and they are especially pernicious. One emerging train of thought is called the over-mineralization theory of aging. It posits that metals like iron build up within cells and increase the rate of free-radical generation. This might explain the negative effect of red meats on lifespan—they contain lots of iron.

Regardless of how they're formed, free radicals can react with just about every component of the cell. Since these reactive oxygen molecules are mostly made inside mitochondria, the first concern is the mitochondrial DNA. This DNA is much less protected than DNA in the nucleus. It sits free within the mitochondria, where it is exposed to attack by free radicals. Also, the DNA repair facilities in the mitochondria are much less sophisticated than those in the nucleus. These factors conspire to make mutations in mitochondrial DNA much more common than with nuclear DNA.

Free radicals can also leak out of the mitochondria and attach themselves to lipids in the cell membrane and proteins in the cytoplasm. Hydrogen peroxide is even stable enough to travel to the nucleus, where it can damage the chromosomes that hold the sacred recipes for all of the cell's thousands of proteins. So while mitochondria provide an incredible boost to the cell's energy capacity, they also pollute the cell with dangerous oxygen-based by-products. Over time, eukaryotic cells evolved powerful anti-oxidant enzyme systems to deal with these threats. An enzyme called superoxide dismutase (SOD) is produced in both mitochondria and the cytoplasm to turn the superoxide free radical into hydrogen peroxide. Other enzyme systems such as catalase can then convert hydrogen peroxide into water.

There was a great deal of excitement roughly twenty years ago about the role of these oxygen-based free radicals in aging and disease. This led to the aggressive marketing of a wide range of "antioxidants" that could purportedly mop up the reactive oxygen molecules with the promise of dramatic improvements in health and lifespan. The results were extremely disappointing. Study after study has shown that such artificial antioxidants have little or no benefit.

Although we don't yet know exactly why antioxidant supplements have been such colossal failures, experts suspect two factors are at play. First, the eukaryotic cell has learned to monitor levels of reactive oxygen species and use them as a signal to know when the cell nucleus needs to make more of its own antioxidants. Taking artificial antioxidants may do more harm than good by interrupting this important form of communication between mitochondria and nuclear DNA. Furthermore, taking high doses of one antioxidant may just increase the levels of an even worse free radical. For example, some antioxidants will help our cells turn superoxide into hydrogen peroxide. But if we don't have enough of the antioxidants to deal with the extra hydrogen peroxide, it may do its own damage. This is dangerous because hydrogen peroxide is more stable than superoxide. Taking antioxidants that turn superoxide into hydrogen peroxide may just create a reactive oxygen species that is better able to travel to the nucleus and attack the DNA there.

Note that the problem with artificial antioxidants does not seem to extend to natural antioxidants found in plants. Study after study has demonstrated the value of eating a variety of brightly colored plants rich in different types of antioxidants. We'll talk more about the specific benefits of plants in chapter 12.

So far we've been focusing on the downside to the union between an archaeon and an alphaproteobacterium. However, there are intrinsic problems that arise with multicellularity as well. This is the second factor thought to be behind the failure of antioxidant supplements.

As we discussed in the chapter describing Life 3.0, multicellular creatures have to struggle with fundamental conflicts of interest. Trillions of cells must subjugate their individual interests in favor of the collective good. They must agree to grow and divide only when explicitly told to do so. Chemical messengers called growth factors signal cells when it's time to divide. It's quite possible for a group of cells to break this compact and to grow and divide even in the absence of the appropriate messages. This can result in the formation of a cancer that can threaten the existence of the entire organism.

Just as reactive oxygen species (ROS) are used as signals within a cell, multicellular creatures learned to employ them to communicate between cells. It's easy to see how a confederation of cells could spot "cheaters" by monitoring their production of ROS, just as nations police international weapons treaties by looking for telltale traces of radiation. A cell running too "hot" might be getting ready to divide when it shouldn't. Once again, high doses of exogenous antioxidants might mask this activity and allow cancer to progress undetected.

Another potential problem lies in the operation of the immune system that multicellular organisms must develop to protect themselves from invaders such as bacteria and viruses. The immune system has an incredibly complex task in trying to distinguish self from non-self. The difference between an animal protein and the corresponding protein in a pathogen can be quite small. When the immune system mistakes its own cells for dangerous invaders, the results can be a disastrous form of friendly fire called autoimmunity. Immune cells use ROS as chemical weapons against bacteria and other invaders. This is yet another reason to be cautious with high doses of antioxidants.

Hopefully this chapter gives you a sense of the tradeoffs involved in moving to an oxygen-based metabolism. The situation is very much like the debate around nuclear power. It's great to be able to produce large amounts of energy by harnessing a dangerous process like radioactive decay, but you have to consider the by-products and treat them very carefully. As we learn repeatedly throughout life, there is no free lunch.

References and Further Reading

N Lane (2005). *Power, Sex, and Suicide: Mitochondria and the Meaning of Life*. Oxford University Press.

Y Liu, G Fiskum, and D Schubert (2002). Generation of reactive oxygen species by the mitochondrial electron transport chain. *Journal of Neurochemistry*, 80(5): 780–87.

JM Son and C Lee (2019). Mitochondria: multifaceted regulators of aging. *BMB Reports*, 52(1): 13–23.

7

Why Do We Age and Die?

"Cells are required to stick precisely to the point. Any ambiguity, any tendency to wander from the matter at hand, will introduce grave hazards for the cells, and even more for the host in which they live... There is a theory that the process of aging may be due to the cumulative effect of imprecision, a gradual degrading of information. It is not a system that allows for deviating."

LEWIS THOMAS

WHY DO we age and die? Humans have pondered this question since time immemorial. Why is it that the perfect beauty of childhood and adolescence soon gives way to dysfunction and eventual death? How do our bodies lose the almost magical balance and resilience of youth? Why does the incidence of all sorts of chronic diseases start to increase rapidly when we get into our forties and fifties?

In the last chapter, we examined the dark side of the improbable union of an archaeon and a bacterium that resulted in the birth of the eukaryotic cell more than two billion years ago. As we've seen, the seeds for our own destruction were planted right then and there. In this chapter, we'll take a look at the aging process and how it works.

What is Aging?

Aging is an area of biology that has largely been ignored until the last few decades. Most people assumed that there wasn't a whole lot to learn. Just as cars and other machines rust and slowly decay, the same thing happens with the human body until it wears out. Why study something so inexorable?

The new millennium has ushered in a major change in attitudes and approaches toward aging. As we've come to understand more about how cells operate at the molecular level, we're starting to realize that there's more to aging than we once thought. While it's true that some aspects of aging represent the accumulation of simple wear and tear over time, many others look suspiciously like "programs" executed by the DNA in our cells. If there are such programs, then it's theoretically possible to alter their behavior or even shut them down entirely. And if we understand the mechanisms behind the various types of wear and tear, we might be able to deal with them as well.

So what is aging? We know it when we see it—or feel it. There are the obvious external manifestations like wrinkled skin, graying hair, saggy muscles, a slowed gait, and so on. However, just as much is going on under the covers. Brain function slows, eyesight and hearing become less acute, organ systems become less effective, and immune function deteriorates to leave us more vulnerable to infection. Aging seems to affect every single operation in our bodies. Are there commonalities across all of these systems and effects?

Aging might best be thought of as a loss of resilience. Consider a brand-new trampoline. You can jump on it as hard as you like. It deforms to compensate for your weight and momentum but snaps right back into its original shape. That is true resilience—the ability to experience a stress, then bounce back to your initial state. We live this when we're younger: We recover from injuries in no time. We fight off infections quickly, sometimes without even noticing that we were sick. Our cuts and bruises heal quickly. Our bodies are incredibly resilient to just about any assault.

Fig. 50: Aging process chart

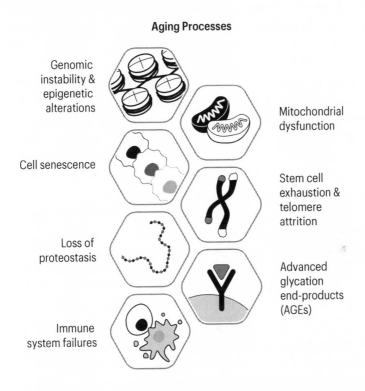

right back into its original shape.

But as we age, the resilience of youth begins to decline. The strained muscle takes much longer to heal. Infections make us sicker, and we don't kick them as quickly. We have more aches and pains, even after relatively minor exertion. We're like an old trampoline that has been left outside, exposed to sun and wind, heat and cold. Our springs have started to rust and some of the fibers making up the fabric of our bed have begun to fray and become brittle. If someone bounces on us, they don't go as high. And if too great a load is placed on us, we simply break in unpredictable ways. A scientist could study us and come up with different names for the various ways we might malfunction. So a spring breaking in the corner might be called Washington's disease, while another breaking along an edge might be named Jefferson's disease. A tear in the center of the bed might be called Adams's disease, while a rip closer to the frame might be labeled Madison's syndrome. Yet they're all just different manifestations of general wear and tear. Exactly which defect (disease) occurs first is largely a matter of chance.

In recent years, scientists have started to examine aging more closely and have created the following framework to better understand the different processes involved.

Let's take a look at the best understood of these processes.

Genomic Instability and Epigenetic Alterations

Inside the nucleus of each of our cells lie twenty-three recipe books written in the four-letter language of DNA (A, T, C, and G). Each of these books contains recipes for hundreds of proteins. For example, the start of the recipe for one component of hemoglobin looks like this:

```
ACTCTTCTGGTCCCCACAGACTCAGAGAGAACCCA
CCATGGTGCTGTCTCCTGCCGACAAGACCAACGTC
```

We read this recipe three letters at a time. Every triplet codes for a specific amino acid—the triplet ACT, for instance, specifies the amino acid threonine. Overall, our library of DNA recipe books contains the secret formulas (genes) for about twenty thousand different proteins. Just to be safe, we actually keep two copies of every book—so forty-six in total. Inside the cell, we call these books *chromosomes*. Each book (or chromosome) is a long string of DNA letters—so long that they would break if they were just left to flop around. Eukaryotic cells came up with an elegant scheme to maintain these long, fragile strands of DNA—most of the strand is wrapped around spools made of proteins called histones. This way, the majority of the DNA is wrapped up in a very compact form that keeps it safe from the many dangerous chemicals in the cell.

Fig. 51: DNA looped around histones

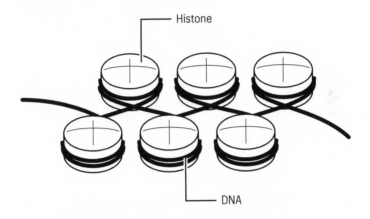

The DNA looped around the histone spools is not very accessible. If the recipe for a particular protein is needed, the cell can slide the spools around to put that recipe in the open area between two of them.

The fact that all fifteen trillion or so of the cells in our bodies have the same DNA content in the same forty-six chromosome recipe books is confusing to many people. If the DNA controls what is going on in a cell, and if all cells have the same DNA, why does a liver cell (hepatocyte) look and behave differently from a brain cell (neuron)? As you'll recall, a new human (and every other animal) starts from a single fertilized egg. This egg is a single cell with twenty-three DNA recipe books that came from the father and another twenty-three that came from the mother. This cell divides to make two, those two divide to make four, and so on. At each division, a decision must be made. Should the two daughter cells be exactly like the cell they came from? Or should the daughter cells specialize in some way? This pattern of division ultimately leads to the different specialized tissues that make up the body.

Fig. 52: Adult blood stem cells/other stem cells

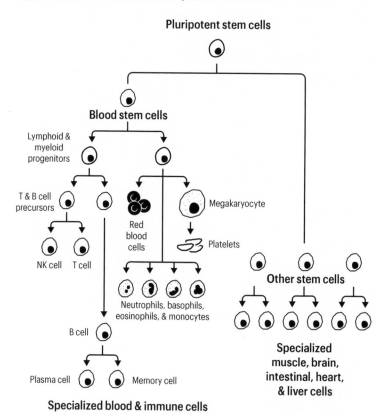

Specialized blood & immune cells

At each branch point along the path to becoming a specialized cell, certain gene recipes must be turned on and others must be turned off. But how do cells do this? The first hypothesis was that cells just deleted genes they didn't need anymore. That concept was proved wrong. Cells don't get rid of any DNA recipes as they specialize, they just mark those recipes in a way that makes them inaccessible. This is the process of *epigenetics*, which means "above the genome."

Cells attach methyl groups (-CH₃) to the DNA corresponding to genes they no longer need. We consider these to be epigenetic marks. Such epigenetic marks flag a DNA recipe as no longer available. One effect is to wrap that section of DNA very tightly around a histone spool so that it can't be used. This is why a skin cell doesn't produce hemoglobin and a neuron doesn't make insulin. The DNA recipes for those proteins have been marked "do not use!" in the chromosomal libraries of those cells. Continuing our analogy comparing DNA to twenty-three pairs of recipe books, think of this as being like putting a clip on the pages of certain recipes so they can't be used.

Fig. 53: Big-picture genetic overview

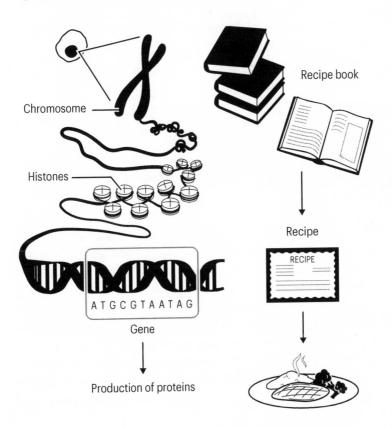

Chromosome

Recipe book

Histones

Recipe

A T G C G T A A T A G

Gene

RECIPE

Production of proteins

But what happens when a skin cell divides to produce two new skin cells? It's important that all of the epigenetic marks in the mother cell be transmitted to both daughter cells. From there, new marks can be made to further specialize one or both of the daughter cells.

So, we have the big picture. The nucleus of the eukaryotic cell stores the DNA in twenty-three pairs of chromosomes. Each chromosome is a long strand of DNA with many sections wrapped around histone spools. A gene is just a section of a DNA strand that contains the recipe (amino acid code) for a particular protein. As cells specialize into specific tissues, certain genes are turned off—many for the rest of the life of that cell. For example, once a cell takes a step down the path to becoming a liver cell, its hemoglobin gene is switched off forever and it will never produce that protein.

This all sounds perfectly elegant. What could go wrong? Unfortunately, a lot. For one thing, our cells are constantly bombarded with radiation, toxic chemicals from pollution, and most importantly the reactive oxygen species produced by our own mitochondria. All of these, and even chemicals from the foods we eat, can react with our DNA to cause the delicate strands to break.

Now don't panic. We have evolved fantastic systems of protein maintenance workers that constantly monitor our DNA for breaks and are amazingly good at fixing them. But no repair system is perfect, and some defects slip through. This is especially true for dividing cells because there's so much work that has to be inspected. With a few exceptions, every cell has a total of 6.4 billion DNA letters in its nucleus—3.2 billion in the twenty-three chromosomes from the mom, and the same number from the dad. When a cell divides, it must copy all of this information using the chemicals that make up DNA. The system is so exquisitely fine-tuned that it makes only about one error in each cell division—fewer than one mistake in each billion DNA letters copied.

But every once in a while, a T gets inserted instead of a G, etc. The genome is so big and the protein-coding recipes so small that the odds of a change adversely impacting the daughter cell are low, but it does happen. After all, our cells have to divide many times to form and then maintain our bodies. If an error slips into the final division that produces a skin cell, no big deal. That cell is going to live only a few weeks before being sloughed off and replaced by a new one. However, the wrong error in a stem cell could be catastrophic since it could affect all the eventual descendants of that cell. As we'll see later, this can even result in cancer.

Changes with Aging

As we age, the occasional errors introduced during cell division or even by day-to-day operations to repair DNA breaks start to add up. Where we really get into trouble is when our maintenance crews start to become less efficient. Now we're at risk of a dangerous positive feedback loop: DNA damage → decreased DNA repair → more DNA damage. Imagine the effect of a mutation in one of the key members of the maintenance crew like the genes BRCA1 and BRCA2. Some people are born with versions of these genes that don't function properly from the beginning. This leaves them at elevated risk of developing certain types of cancer because their DNA repair is deficient.

In addition to errors (mutations) in the DNA sequence itself, aging causes our epigenetic machinery to slow down too. The DNA doesn't stay as tightly coiled around the histone spools. Some of the methylation marks that were supposed to make certain genes unavailable start to fall off. In young bodies, the process of reading DNA and manufacturing the proteins it specifies is very precise. Everything is very binary—a gene is either on or off—and we say the entire system is "quiet." The only things happening are those that are supposed to be happening. As we age, our cells become "noisy." Some genes that were previously turned off start to operate part of the time, producing small levels of unneeded proteins. This wastes energy that now can't be devoted to finding and fixing problems. When the situation gets too far out of control, the cell commits suicide. Over time, our tissues have fewer healthy cells to carry out the job of that tissue. We can easily see the effect under the microscope or by looking at our own skin.

Fig. 54: Cellular structure of old and young tissue

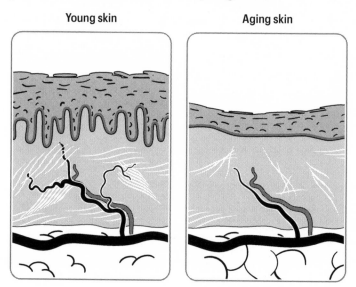

Mitochondrial Dysfunction
....

Mitochondria have long been suspected as culprits in the mystery of aging. As we discussed in the last chapter, mitochondria play with fire in the form of molecular oxygen. They are mostly able to keep it under control, but every once in a while, oxygen grabs an electron when it shouldn't and becomes a dangerously reactive free radical (or reactive oxygen species). Unlike the host cell that keeps its DNA encased in a protective membrane (the nucleus), the mitochondrion's DNA floats free in the matrix, where it is exposed to these potential oxygen-based assaults.

You see, free radicals just love to react with DNA. In the process, they can garble the sense of the recipes contained in the DNA and cause the mitochondria to produce worthless proteins. Free radicals can also attack proteins and lipids in the mitochondrial membranes. This whole process reduces the effectiveness of mitochondria and decreases their production of ATP. In young, healthy cells, these defective mitochondria are tagged for destruction and quickly incinerated in lysosomes—little acid-filled bags within the cell. However, this recycling of old, damaged mitochondria becomes less effective with age. The result is a buildup within cells of lethargic mitochondria, which are less able to carry out their assigned tasks. Tissue samples from older people show cells with fewer and fewer good mitochondria as they age. The defective mitochondria also become leaky, allowing free radicals to spill out into the host cell and travel to the nucleus, where they can wreak havoc with the cell's central DNA.

Cell Senescence

It shouldn't come as a big surprise that evolution has developed a scheme to prevent cells from dividing when they shouldn't. For single-celled organisms, the decision to divide is fairly easy. If there's sufficient nutrition in the area and no other major stressors, first copy all the components that will be necessary for two cells, then go ahead and divide.

The situation became much more complex when organisms became multicellular. Allowing each cell to make the decision to divide on its own would be disastrous. Multicellular organisms developed a wide range of signals and safeguards to control exactly when each cell divides. Cells have protein receptors in their membranes that protrude into the outside world like antennae listening for signals. Those signals come in the form of chemical messengers, so if, for instance, you cut your skin, your body would release growth factors that tell stem and progenitor cells in the skin that they need to divide to heal the wound.

But what if a cell has suffered damage that would make it dangerous to divide? For example, what if a cell has been exposed to radiation and its DNA has suffered numerous breaks that need to be repaired? Trying to divide under these circumstances would be catastrophic. If the division succeeded at all, it could produce a daughter cell with serious DNA defects that could result in a cancer.

It turns out that cells have elaborate quality control mechanisms that allow them to sense different kinds of damage. Cells can even sense that they are being driven too fast, a feature of many cancers that push an aggressive agenda of growth and division. When the quality control machinery of the cell determines that something is so wrong that it would be dangerous to divide, it can revoke the division permit—kind of like a factory worker who can shut down the entire assembly line in the event of a problem.

Although the cell has many quality control systems, we'll focus on two important proteins: p21 and p16. Both can be generated in response to a problem in the cell such as DNA damage, short telomeres (see below), high levels of oxidation, or an overly aggressive rate of division. Less severe damage instigates the production of p21 and invokes only a temporary shutdown, giving the cell enough time to repair the damage and then move forward with division. When the detected problem is so severe that there's no hope for recovery, p16 is made. In essence, p16 hangs a "do not divide!" sign on the cell. We call this state in which the cell's division permit has been permanently revoked *senescence*.

Cell senescence stops damaged cells from reproducing. This might reduce the risk of cancer. When we're younger, we have very few senescent cells, and our immune system recognizes and eliminates them easily. As we age, however, senescent cells begin to accumulate, and our immune system no longer clears them as efficiently. The problem is that these senescent cells are still metabolically active but can no longer divide. They're like a car with a driver who's pressing the accelerator and the brake pedal at the same time. The car can't go anywhere, but it spews out exhaust fumes and makes a heck of a racket. Like the car, senescent cells secrete toxic chemicals that pollute the local environment. They also produce proteins called cytokines that increase the level of inflammation. These cytokines and other pollutants can damage nearby cells enough to cause them to become senescent as well.

Senescence is especially damaging when it hits a stem cell. Recall that stem cells sit in every tissue of the body and serve as reservoirs of replacement cells in response to damage. If a group of liver cells dies after a night of heavy drinking, a liver stem cell is spurred into action to divide and replace the fallen comrades. But if such a stem cell becomes senescent, it can no longer do its job. The tissue in which it sits gradually goes from being a dense collection of healthy cells to a few normal cells surrounded by fibrotic (scar) tissue and mostly useless senescent cells. In this way, senescence contributes to another hallmark of aging—stem cell exhaustion. As we'll see later, a promising area of aging research is focused on killing off senescent cells using drugs known as senolytics.

Fig. 55: Senescent cell: age-associated disease

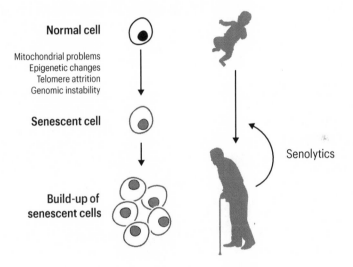

Stem Cell Exhaustion and Telomere Attrition

Stem cells are special for two reasons. First, they are self-renewing. This means they can divide into two cells: a stem cell and a more specialized cell. As shown in figure 56, often a stem cell will create another "lieutenant" cell that will actually do most of the dividing. The lieutenant cell is called a *transit amplifying* or *progenitor* cell. These are the cells that will gradually populate a tissue with fully differentiated cells—like neurons in the brain or muscle cells in the biceps. This scheme allows the stem cells at the top of the hierarchy to divide only occasionally. This is important because cell division is very dangerous for the DNA of a cell. The entire set of DNA recipe books must be opened up and copied during cell division. The fragile strands of DNA lie exposed to all sorts of chemical and physical hazards and are subject to dangerous breaks that must be accurately repaired to prevent mutations.

Fig. 56: Stem cell to mature cell

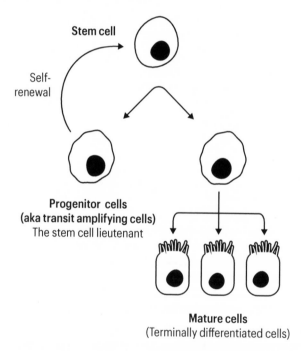

It may seem that this design would allow tissues to grow and regenerate forever. Unfortunately, that's not the case. Each cell has a repeating section of DNA that forms the endcaps of its chromosomes—the telomeres. When the chromosome is copied during cell division, this telomere gets a tiny bit shorter. When the endcap is gone, the chromosome unravels and the cell self-destructs. This means that mammalian cells can divide only fifty to sixty times over the course of their lives. When a stem cell dies, it is lost forever and can't be replaced.

Each tissue around the body has a small reservoir of stem cells. Some tissues have more because those tissues are constantly growing and turning over their differentiated cells. For example, cells in skin tissue and intestinal lining live only a few days and must constantly be replaced. Other tissues like the brain and heart are comprised of long-lived cells that may have to last one hundred years or more. They contain very few stem cells. For this reason, damage to the brain or heart is very difficult, often impossible, to repair.

When a stem cell in a tissue dies, that part of the tissue is now incapable of replacing lost cells. As the cells downstream of the missing stem cell die, they are replaced by scar tissue and fat. We can see this happening in our skin and other tissues as we age. As we slowly lose functional cells, the tissue becomes less and less resilient and increasingly frail.

Loss of Proteostasis

Hopefully by now you appreciate the importance of proteins. They do all the real work within the cell. Of course, DNA is critical in order to store the recipe book for all the proteins the cell needs to make. Lipids (fats) form the enclosing membrane of the cell and all its internal organelles like the nucleus and mitochondria. Carbohydrates provide energy and even act as structural components in some cells. But proteins are the real marvels. They serve as catalysts to speed up important chemical reactions many thousandfold; reactions that would otherwise take years are complete in a few seconds. Proteins also form channels to allow the selective import or export of certain molecules (e.g., waste lactate during exercise) that wouldn't otherwise be able to traverse the cell membrane. Proteins even make tiny machines like the ATP synthase that utilizes the power from protons passing through to squeeze a phosphate onto a molecule of ADP.

We can think of proteins as strings of amino acids, kind of like beads on a thread where there are twenty different types of beads. However, this is where the analogy breaks down. Proteins don't just lie around in the cell as floppy chains of amino acids. Instead, they fold up into complex shapes like the bent pieces of metal we use to make functional machines.

Fig. 57: Folded and unfolded amino acids

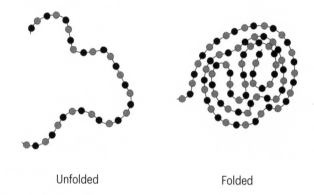

Unfolded   Folded

The shape a protein takes is dictated by its exact amino acid composition. For example, amino acids such as arginine carry a positive charge. Others, such as aspartate, carry a negative charge. It's easy to see how a protein with an arginine in one position and an aspartate in another might fold to bring the positive charge into contact with the negative. In general, proteins fold into a shape that represents the one with the lowest energy, just like a ball at the top of a flight of stairs will tend to roll all the way down once pushed.

In addition to trying to get amino acids with opposite charges close to each other, proteins leverage another type of interaction— the one between amino acids and water. Some amino acids are hydrophilic—they love water and are happy being in an aqueous environment. The charged amino acids are among these. Other amino acids, however, are hydrophobic and will go to great efforts to get away from the stuff. Because proteins have to work in the watery environment of the cell, their outsides are surrounded by water. Just as olive oil will slowly separate from water in salad dressing, the hydrophobic amino acids within a protein prefer to bunch up together on the inside of a protein where they can avoid H_2O. Thus the interior of a protein tends to be a sticky, oily place much different from the hydrophilic amino acids we see on the outside.

But even protein folding can go wrong sometimes. That's why cells have evolved a set of repair proteins that look for misfolded proteins and help them adopt the right shape. These repair proteins are called *chaperones* and are critically important to the proper function of the cell. We'll talk more about them later.

Ordinarily, the lifecycle of a protein looks something like this:

1   The cell decides that it wants to make a protein.

2   The cell locates the stretch of DNA that specifies the recipe (amino acid sequence) for that protein.

3   If that stretch of DNA is wound around a histone spool, that DNA is loosened from the spool. The spool might also be moved in order to better free up the needed DNA.

4   The DNA sequence containing the recipe for the protein is copied into RNA.

5   The RNA leaves the nucleus and enters the cytoplasm of the cell.

6   A ribosome is assembled on the RNA.

7   The protein specified by the RNA is constructed.

8   If necessary, a chaperone helps the newly constructed protein fold into the right shape.

9   The protein does its work within the cell.

10 When the protein is no longer needed—or if it becomes unfolded—it is tagged with a "recycle" label.

11 The tagged protein is taken to a garbage disposal called the proteasome, where it is chopped up into amino acids for reuse.

As we age, many parts of this elaborate system start to malfunction. There may not be as many chaperones around as needed to help proteins fold. Or the recycling system can become less efficient and have a hard time keeping up with the volume of misfolded proteins. Chaperones can become overwhelmed trying to deal with the heap of misfolded proteins and have less time to do their normal job—helping newly synthesized proteins fold correctly.

Our bodies seem to produce more damaged and misfolded proteins as we age. The oxidative damage we've discussed previously contributes to the increase in misfolded proteins. Figure 58 shows results from a roundworm, but the same thing happens to all animals, including us.

Fig. 58: Adult age/protein fraction

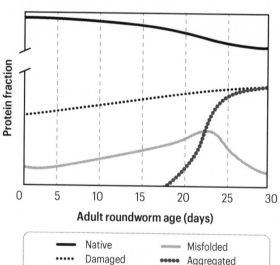

Adapted from Santra et al (2019).

Notable in this diagram are the lines labeled *Misfolded* and *Aggregated*. When too many misfolded proteins accumulate within a cell, the danger of aggregation increases dramatically. As we've said, misfolded proteins no longer keep their "sticky" hydrophobic parts on the interior of a nicely constructed ball or other shape. Now those parts are exposed, making it possible for the sticky part of one protein to adhere to the sticky part of another. This can cause the rapid accumulation of clumps of misfolded proteins that can result in fibrils, tangles, clusters, and other complex shapes. As we'll see in the next chapter, this accumulation of sticky proteins can lead to cell death and result in disorders such as Parkinson's disease. Increasing evidence demonstrates that protein aggregation contributes to the chronic inflammation seen in aging in general, and particularly in neurodegenerative diseases like Alzheimer's, Parkinson's, and Huntington's.

Advanced Glycation End-Products (AGEs)

You know that nice glaze that develops when you put barbecue sauce on meat and cook at high temperature on a grill? Unfortunately, something quite similar happens in our cells too. Sugars such as glucose can bind to proteins and membrane lipids to create chemical compounds called advanced glycation end-products: AGEs for short, fittingly. This contributes to protein misfolding and also to decreased membrane flexibility. AGEs build up over time and can easily be measured in skin. The problem is especially serious in people with diabetes because they often have high levels of glucose in their blood. AGEs can increase the baseline level of inflammation in cells and contribute to the increased level of inflammation frequently seen with advancing age.

Fig. 59: AGEs' impact on body

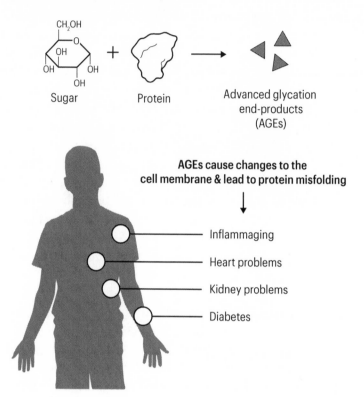

Sugar + Protein → Advanced glycation
end-products
(AGEs)

**AGEs cause changes to the
cell membrane & lead to protein misfolding**

Inflammaging

Heart problems

Kidney problems

Diabetes

As illustrated in figure 59, AGEs have wide-ranging impact throughout the body. They may well explain high rates of cancer among people who adopt high-carb or high-fat diets. AGEs also increase the overall level of oxidation within the cell, accelerating problems with misfolded proteins and subsequent protein aggregates.

Immune System Failures

We're appreciating the importance of our immune system more with each passing day. Until recently, it was thought to do little more than fight off pathogens such as bacteria and viruses. Now we know that the immune system does far more and is a major factor in our day-to-day health, including keeping cancer at bay.

Just as the American military is organized into different services (Army, Navy, Air Force, etc.), the immune system contains two main branches: myeloid and lymphoid. Both generate various types of "white cells," all of which come from a special type of progenitor called a hematopoietic stem cell (HSC) that resides in the bone marrow.

Fig. 60: Bone marrow HSC

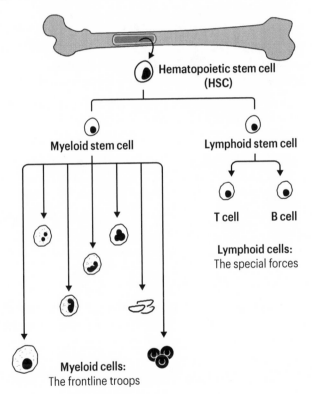

We can think of the cells in the myeloid "service" as simple infantry soldiers. They don't go through much training and are taught to look for basic features common across many different types of invaders. For our purposes, the most important myeloid cells are neutrophils and monocytes. They are the frontline troops that are sent in at the first sign of trouble. In general, they shoot first and ask questions later. While they are effective at slowing down invaders, they tend to be somewhat indiscriminate and create lots of collateral damage to our own cells. The rapid-response process they participate in is called *inflammation*. When they arrive at a trouble site, they quickly release chemicals called cytokines that open up local blood vessels to allow more first responders to enter the area. The side effect of this response is pain, swelling, and heat—the sort of inflammation we all know from even a simple splinter.

The lymphoid branch of the immune system is much newer from an evolutionary standpoint. It is usually also much smarter and more discriminating. We can think of the T and B cells that make up the lymphoid branch as highly trained special forces. They are summoned only when the initial myeloid response cannot deal with the problem. T and B cells are exquisitely sensitive and trained to look for a very specific type of threat, much as a small special forces team can be sent in to take out a particular target.

In our youth, we have a roughly fifty-fifty balance between the myeloid and lymphoid branches of our immune system. As we age, however, this balance becomes skewed in favor of the infantry. We have fewer and fewer lymphoid cells, so the myeloid branch comes to dominate. This results in a condition called *inflammaging*—a generalized increase in inflammation throughout our bodies. It's as if there's a group of twenty-year-old soldiers on every corner, armed to the teeth. Although they're nominally there to protect us, it's easy for them to overreact to even minor incidents, and chaos can ensue. If even slightly provoked, they can turn their weapons on normal citizens and do more harm than good. This is what we see in autoimmune diseases such as lupus and rheumatoid arthritis.

One important recent discovery is that the lymphoid arm of the immune system is tremendously important in fighting cancer. It turns out that our bodies are generating new cancer cells every day. Fortunately, our lymphoid cells are very adept at recognizing and killing these rogue cells. But as that part of our immune system starts to degrade with age, so does our cancer-fighting ability. This whole situation is exacerbated by the fact that the thymus, the gland where T cells are trained to attack only invaders and not our own tissues, withers away as we age. So we have the double whammy of fewer lymphoid cells with less ability to differentiate between foreign invaders and our own components.

NOW YOU understand some of the important factors that underlie aging and the development of several chronic diseases. In the next chapter, we'll examine some of these diseases more closely.

References and Further Reading

M Borghesan et al (2020). A senescence-centric view of aging: implications for longevity and disease. *Trends in Cell Biology*, 30(10): 777–91.

C Lopez-Otin, MA Blasco, L Partridge, M Serrano, and G Kroemer (2013). The hallmarks of aging. *Cell*, 153(6): 1194–1217.

C Ott et al (2014). Role of advanced glycation end products in cellular signaling. *Redox Biology*, 2: 411–29.

M Santra et al (2019). Proteostasis collapse is a driver of cell aging and death. *PNAS*, 116(44): 22173–78.

DP Turner (2017). The role of advanced glycation end-products in cancer disparity. *Advances in Cancer Research*, 113: 1–22.

8

Diseases of Aging

I N MEDICAL school, I dutifully learned about a mind-numbing number and variety of diseases: Alzheimer's, Parkinson's, cancer, diabetes, heart disease, and so on. Each was presented as its own unique disorder with separate signs, symptoms, and causes. I came to imagine life as a field across which you had to run. The various diseases were like different camouflaged pits that you had to avoid falling into at all costs, each one an isolated, completely independent danger. If you could just make it to the other side of the field without falling into one of the pits, you would live a long and healthy life. I learned further that some of the pits were easier to avoid than others because of external associations that could tip you off to their location. For example, growing next to diabetes pits were usually hamburger trees. Don't go near a hamburger tree and you won't fall into the diabetes pit. Parkinson's pits were frequently found near pesticide bushes. Keep away from those bushes and significantly reduce your chance of tumbling into a Parkinson's pit. I'm sure that most of my colleagues ended their medical studies with the same impression. All these diseases could be traced to a multitude of different underlying causes and were mostly independent of each other.

But if this is really the case, why is it that the incidence of so many diseases starts to rise exponentially as we move past middle age? Consider the following chart from the American Association for the Advancement of Science (AAAS) showing the incidence of cancer, heart disease, infectious disease, and Alzheimer's with age.

Fig. 61: AAAS chart

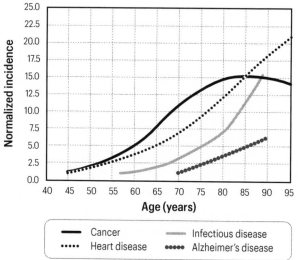

Adapted from Rae et al (2010).

As we saw in the previous chapter, there are many changes that occur with age that collectively decrease the body's ability to deal with various kinds of stress. Not a bad day at work kind of stress but loads analogous to a person jumping on a trampoline. And unlike my impression from medical school, these diseases aren't independent at all. They represent different external manifestations of the various functional declines we discussed in the

previous chapter: stem cell exhaustion, mitochondrial dysfunction, etc. Going back to the field with a bunch of camouflaged pits,
a better metaphor would be a field overlying ground that has been
hollowed out by a giant colony of moles. Yes, there might be telltale signs of burrowing activity on the surface that could allow you
to avoid stepping on an especially weak part of the ground. But
the important point is that there's no real mechanistic difference
among the pits. They are all interconnected and all caused by the
same underlying factor—the hollowing out of the ground by the
moles. Exactly where you fall in is largely a matter of chance. One
person might hit the Alzheimer's pit while another disappears
down the diabetes hole, but that distinction is mostly meaning-
less. (As a quick aside, some forms of Alzheimer's are now being
called type 3 diabetes.) So, overall, a person's resilience starts to
decrease as they reach their mid-forties. This decline happens in
all tissues throughout the person's body but can be mitigated by
healthy lifestyle choices. Let's now look at some of the most common diseases of aging and their specific characteristics.

Diabetes

Type 2 diabetes has reached epidemic proportions and now afflicts
one in ten of all Americans (including children). Over half of Americans are either overtly diabetic or prediabetic. The prevalence
increases rapidly as people move beyond their mid-forties.

The overall course of type 2 diabetes is relatively easy to understand. The two main players are glucose and insulin. Glucose is
the most common form of sugar used for energy by the body. As
food is absorbed by the gut, it is broken down into individual components that flow into the blood. Carbohydrates mainly break
down into glucose, the primary energy source for muscles, brain,
and other energy-intensive tissues. But glucose requires a "door"
to be opened in the membranes of these cells in order to enter.
This door comes in the form of a protein that transports glucose

from the bloodstream to the interior of the cell where it can be used for energy. Cells open these doors (put the protein transporters into the cell membrane) only when a protein hormone called insulin tells them to.

Insulin is manufactured by special cells in the pancreas called *beta cells*. When the pancreas sees levels of glucose rising in the blood, it releases insulin into the bloodstream. Insulin quickly flows around the body telling cells to open their doors and take in the available glucose. What happens if the doors don't open? Type 2 diabetes.

Fig. 62: Natural history of type 2 diabetes

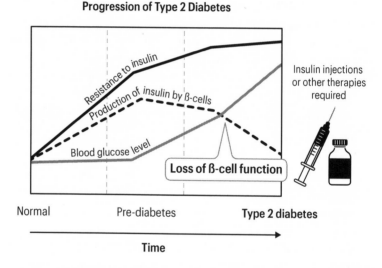

Type 2 diabetes is often accompanied by obesity. A simplistic way to think about the disease is this: When you start taking in more calories than you need, your pancreas must produce more insulin to get muscle and other cells to open their doors. However, these cells can only burn so much glucose, so they begin to store the excess as fat. Fat is fine if it stays where it belongs—in fat cells. But as fat begins to deposit ectopically (in the wrong places) in muscle, bone, kidney, liver, and other organs, it builds up right within the cells of those tissues. This misplaced fat causes those cells to not want to open their doors when insulin comes knocking. The body tries to compensate by producing even more insulin to knock louder. This strategy works for a while, but eventually muscle cells and those in other organs refuse to listen—they become insulin resistant. At this point, the glucose level in the blood begins to rise and type 2 diabetes sets in. Eventually, the beta cells in the pancreas burn out and die from their frenzied efforts to produce more insulin. Soon, the patient requires insulin injections or other therapies to maintain normal blood sugar levels.

Recall that inflammation is a result of overactivity of the myeloid branch of the immune system—the poorly trained foot soldiers that are ready to attack at the drop of a hat. Inflammation plays a large part in diabetes, as it does in almost every disease of aging. The effects of inflammation are felt in every cell of the body as the result of inflammatory chemicals such as cytokines that circulate in the bloodstream. These chemicals don't just amp up the cells of the immune system but also cause other cells to generate more reactive oxygen species (ROS). The oxidation of lipids and proteins that accompanies inflammation increases the tendency of cells to become insulin resistant.

Type 2 diabetes is largely preventable and can even be reversed in its early stages. Losing excess weight, exercising more, and limiting simple carbohydrates in the diet all can reduce the probability of developing diabetes, and help reverse it once established.

Insulin binds to the insulin receptor, a membrane protein that exists on most cells but especially those in muscle, liver, and fat tissue. This signal activates two different pathways within the cell. One pathway is related to metabolism and tells the cell to open its glucose door by inserting glucose transporters into the cell membrane. This is the pathway that is responsible for using glucose to generate energy. The second pathway is one involved in cell growth and division. We can think of insulin as kind of a "Party time!" signal to cells. It indicates that nutrition is plentiful and it's time to grow—and in many cases to divide. In other words, this part of the insulin signaling is important during growth and is beloved by cancer cells. It takes higher levels of insulin to activate this part of the pathway, which explains in part why people with diabetes and insulin resistance are more prone to cancer. All their cells are receiving a message from insulin to grow and divide, and cancer cells are only too happy to oblige.

Cardiovascular Disease (Heart Disease)

Cardiovascular disease is one area in which huge progress has been made the last few decades. It includes not only the sort of artery blockage that leads to heart attacks but also stroke and high blood pressure. Like type 2 diabetes, cardiovascular disease has a large lifestyle component. Factors such as smoking, lack of exercise, poor nutrition, chronic stress, and many others play a starring role in both the initiation and the progression of the disease. Still, heart disease remains the number one killer in most of the developed world. More than a quarter of people with cardiovascular disease don't even know they have it.

Although there's a genetic component, we can think of cardiovascular disease as being driven by inflammation. Not too long ago, all attention was on cholesterol as the culprit in atherosclerosis (narrowing of arteries with various debris) and cardiovascular disease in general. We now realize that it is cholesterol in the context of inflammation that really does the damage.

When the cells lining the inside of blood vessels (endothelial cells) become inflamed, they insert adhesion proteins into their membranes that act like Velcro. Certain white cells of the immune system attach to this Velcro to slow down at the site of inflammation and squeeze between endothelial cells to enter the tissue space that surrounds the blood vessel lining. There, these white cells turn into a cleanup crew—cells called macrophages. They attempt to eat up the oxidized cholesterol and other junk they find in that area. If there's too much, they bloat with the fatty deposits they've gobbled up and gain a somewhat foamy appearance under the microscope, leading them to be called foam cells. Like people who stuff themselves over the holidays, foam cells become lethargic and lie around. Eventually, these polluted macrophages die and explode, filling the area with even more debris. This debris causes more inflammation, attracts more white cells, and the whole situation spirals out of control. Over time, the inflamed area with dead and dying macrophages swells and bulges into the blood vessel, disturbing the flow of blood. The body attempts to wall it off with a fibrous cap, but this eventually ruptures and creates a clot (thrombus) within the blood vessel.

Fig. 63: Vascular muscle cells

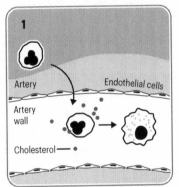
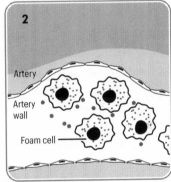
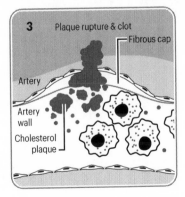

Such a clot can break off and travel through the circulatory system, sometimes getting stuck at a branch point and blocking the rest of the blood vessel. The tissue normally supplied by that blood vessel is suddenly deprived of oxygen and dies. If the blockage occurs in one of the coronary arteries that bring blood to the wall of the heart, a *myocardial infarction*, or heart attack, can result. Sometimes clots travel to the brain and lead to a stroke. Other times they become lodged in the lungs and cause sudden shortness of breath.

As mentioned, inflammation plays a huge role in this whole process. We talked about oxidative damage in the last chapter, and here we see it in action. Cholesterol is transported in the blood by a protein called LDL (low-density lipoprotein). This protein-cholesterol bundle can undergo oxidative damage by any of a variety of means—smoking, mitochondrial dysfunction, etc. The normal diet in most Western countries, with its high levels of animal fat, sugar, and other simple carbohydrates, is also highly inflammatory. In contrast, most plants are anti-inflammatory. Since Western diets tend to be low in fruits and vegetables, the overall balance skews toward inflammation. Inflammation attracts immune cells such as neutrophils and macrophages that use reactive oxygen molecules to fight off invaders, thereby increasing the oxidative damage. Physicians are increasingly looking at markers of inflammation such as C-reactive protein (CRP) in growing recognition that they need to not only bring down levels of LDL and cholesterol but also reduce the level of inflammation in a patient's body. A recent CANTOS study demonstrated that reducing inflammation has a significantly positive effect on cardiovascular disease.

Alzheimer's Disease

Alzheimer's disease is a gradual form of neurodegeneration first described by Alois Alzheimer in 1901. This disorder has grown from a relatively rare condition to one of the major health emergencies of the modern world. It afflicts well over thirty million people around the globe, with numbers growing quickly. Two-thirds of all cases of dementia are due to Alzheimer's.

Alzheimer's is most commonly known for the cognitive deficits it causes, ultimately resulting in dementia. There is still a great deal of controversy around the exact mechanisms of this disease, but focus has centered on the toxic buildup of two types of proteins in the brain: amyloid beta and tau. The association of these proteins with the disease is clear, but less certain is whether their buildup represents cause or effect. That is, perhaps some other problem initiates Alzheimer's, leading to the accumulation of these proteins. Such accumulation could even represent the brain's attempt to compensate for the underlying problem.

The accumulation of amyloid beta is similar to the situation with the protein alpha-synuclein in Parkinson's disease, discussed below. A certain protein starts to misfold, the misfolded proteins begin to clump together, and the clumps grow to form fibrils and then clusters of plaque, as shown in the diagram below.

Fig. 64: Amyloid beta plaque formation

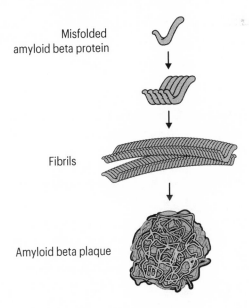

Misfolded
amyloid beta protein

Fibrils

Amyloid beta plaque

This toxic protein aggregation is caused by the improper processing of a normal protein called amyloid precursor protein (APP). As shown below, APP is a transmembrane protein. It extends all the way through the membrane of a neuron and can be thought of as having three sections—outside (extracellular), inside (intracellular), and transversing (transmembrane). Like all proteins, APP is periodically recycled by the cell. The first step in this recycling is for the protein to be snipped in the middle of its transmembrane section by an enzyme called alpha secretase. The two halves can then be disposed of by the usual means.

Fig. 65: PSEN1 PSEN2

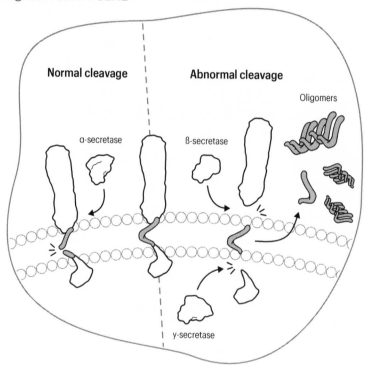

Sometimes, however, APP is cut by a different enzyme called beta secretase. In this case, the cut is where the extracellular section meets the transmembrane section. Another enzyme (gamma secretase) then snips off the intracellular section. These actions leave intact the transmembrane portion of APP. Remember that the inside of the membrane itself is composed of the backbones of fatty acids and is very hydrophobic. The transmembrane parts of most proteins are also hydrophobic. The cutting actions of beta and gamma secretase thus result in a protein section that is very hydrophobic and "sticky." As these sticky protein fragments grow in number, they start to adhere together in the watery environment outside the cell. We call these fragments and their clumps amyloid beta. As these clumps build up outside a neuron, they will eventually kill it, along with other nearby cells.

The "amyloid hypothesis" dominated Alzheimer's research for many years, and billions of dollars were spent developing and testing drugs that could break down amyloid deposits. However, clinical trials have been disappointing with no clear indication that removing amyloid deposits has much therapeutic benefit.

Fig. 66: Tau

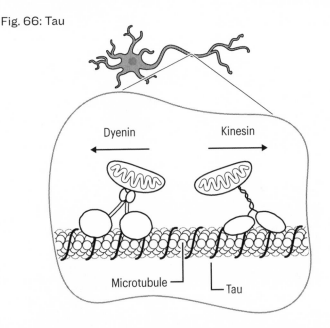

The second protein implicated in Alzheimer's is called tau. Neurons can be very long and have high energy requirements. For this reason, it's critical for them to have a way of moving mitochondria around to bring them where they're needed. The neuron relies on a system of protein cables called microtubules to serve as a scaffold along which mitochondria and other components can be moved around the cell.

Tau is a small protein that serves as a clamp to maintain each microtubule as a taut cable. If tau falls off, the microtubule can disintegrate quickly. Other proteins called kinases can be used to attach phosphate groups to tau. When tau has many of these phosphates, it drops off the microtubules and can then start to form filaments and tangles much in the same way amyloid beta does. Whereas amyloid beta accumulates outside the neuron, tau tangles build up inside. Both result in the eventual death of the neuron. Note that loss of tau from microtubules also reduces their stability and interferes with the ability of mitochondria and other cargo to move where they're needed.

Fig. 67: Alzheimer's sequence

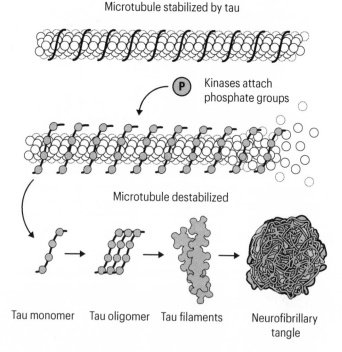

Microtubule stabilized by tau

Kinases attach phosphate groups

Microtubule destabilized

Tau monomer   Tau oligomer   Tau filaments   Neurofibrillary tangle

The current view of the sequence of events in Alzheimer's is reflected in figure 68. The process begins with the accumulation of amyloid beta, which disrupts which synapses—the connections between neurons. Later, tau becomes dysfunctional and leads to the tangles discussed above. These changes provoke alterations in the structure of the brain, eventually leading to memory loss and other cognitive deficits. For reasons that aren't fully clear, damage seems to be especially pronounced in the parts of the brain that control the formation of new memories (the hippocampus) and higher-level executive function (the prefrontal cortex).

Fig. 68: Alzheimer's disease stage

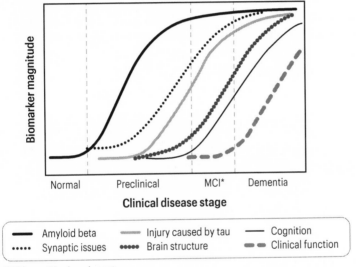

*Mild cognitive impairment

Adapted from Sperling (2011).

An interesting recent development is the notion of Alzheimer's as a form of "brain diabetes" or "type 3 diabetes." Roughly 80 percent of Alzheimer's patients are also diabetic or suffer from insulin resistance. Neurons in these patients often demonstrate insulin resistance as well. The inability for insulin to signal to neurons the availability of glucose in the blood has broad impact.

You will recall that we can think of insulin as a signal to cells that times are good—that there is sufficient nutrition for them to grow. Neurons have very high metabolic demands and consider long-term lack of insulin signaling as a sign that they need to dial back their investments. So they begin to prune their energy-draining connections (synapses) to other neurons. Unfortunately, these connections play a huge role in memory and other cognitive functions. As these connections slowly wither away, memories are lost, along with the ability to perform various reasoning tasks. Thankfully, there's hope that lifestyle changes can reverse the insulin resistance and allow neurons to once again make new connections.

Finally, it's clear that there's a large immune component to Alzheimer's. Patients seem to have overall high levels of inflammation in their brains, perhaps in response to some of the changes described above. Although neurons get all the publicity when it comes to the brain, they are outnumbered by support cells called glia. Some of these glial cells serve an immune function similar to the white cells of the immune system. It appears that extracellular amyloid beta as well as dead and dying neurons evoke an inflammatory response from the glial cells and other parts of the immune system in ways that worsen the underlying damage.

Parkinson's Disease

Parkinson's is a progressive degenerative disorder of the nervous system that typically impacts motor (movement-controlling) areas in the brain. The classic symptoms are tremors, slow movement, intermittent rigidity, and other motor problems. Roughly 1 to 3 percent of people over the age of 65 suffer from Parkinson's. In its later stages, the symptoms evolve to include dementia, loss of smell (anosmia), and a wide range of intestinal problems. Sleep disorders are common, too, along with depression and a reduced range of facial expressions.

The disease usually starts in a small area of the brain called the substantia nigra, in which neurons use dopamine as a neurotransmitter to "talk" to each other. Under the microscope, it's possible to see tangles of proteins consisting mainly of a type called alpha-synuclein inside affected neurons. These can often form large clumps called Lewy bodies.

In addition to these tangles of improperly folded proteins, neurons in Parkinson's patients frequently show mitochondrial dysfunction, especially problems with complex I of the electron transport chain. Drugs like rotenone and MPTP inhibit this same protein complex and can cause symptoms very similar to Parkinson's. In the 1980s, doctors were surprised to see an outbreak of a disorder much like Parkinson's in young recreational users of MPTP. They finally traced it down to MPTP's inhibition of the first complex in the electron transport chain in mitochondria.

You can imagine the impact of disrupting the electron transport chain right at the first step. ATP production by mitochondria falls dramatically. And since electrons can't flow through the chain to complex IV where they're carefully joined to oxygen to make water, electrons leak out all over the place. Some combine with oxygen in a less careful way that generates the reactive oxygen species we discussed previously, further damaging the mitochondria and the neurons that house them. This process is probably taking place in cells all over the body, but it shows up first in the cells that have the highest energy demands—neurons in general and dopaminergic neurons in particular. As we can see from the following diagram, the dysfunctional mitochondria are the common link in the various causative chains that can lead to Parkinson's. This helps explain why the same disease can result from both environmental assaults (poisons, drugs, etc.) and genetic defects in various processes related to mitochondria. For example, there has long been an association between Parkinson's and exposure to pesticides; farmers and people who live near farms have higher rates of the disease.

Fig. 69: Toxin vs genetic models

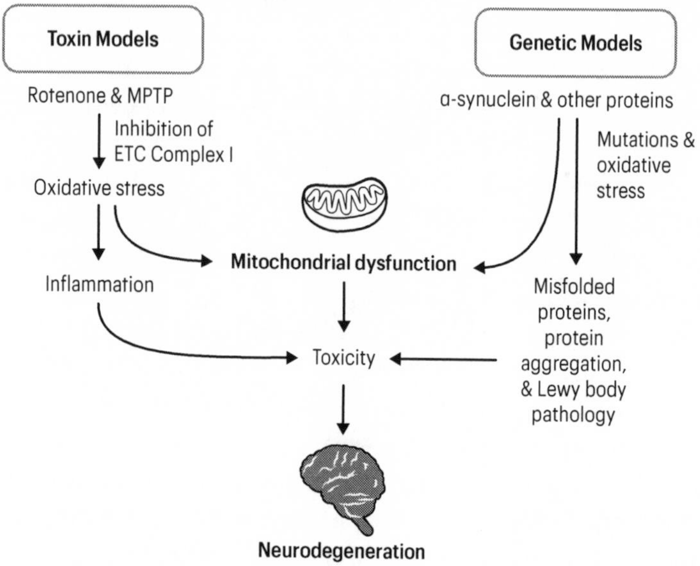

As noted previously, drugs such as rotenone and MPTP inhibit complex I in the electron transport chain, which leads to a decrease in ATP production as well as an increase in oxidative stress. However, failure to clear misfolded proteins like alpha-synuclein as well as general defects in recycling old mitochondria result in the same condition.

Remember from the last chapter that misfolded proteins can sometimes expose sticky parts that are normally kept on the inside of the protein away from contact with water. Once exposed, the sticky part of one protein can adhere to the sticky part of another to create an aggregate, or clump. Certain variants of the alpha-synuclein protein seem to be more predisposed to misfolding than others. Oxidative stress can also contribute to misfolding.

The images in figure 70 illustrate the process well. Normal alpha-synuclein is shown as the simple shape on the upper left with the two graceful, curled legs of the U. These are called alpha-helices and are reflective of a properly folded protein. The misfolded protein is shown to the right of the normal one with the sticky areas shaded. You can follow the arrows in the diagram to see how multiple misfolded alpha-synuclein molecules start to aggregate and eventually form what can only be called a beautiful fibril structure. These fibrils eventually result in the clumps called Lewy bodies that are the classic microscopic feature of Parkinson's.

Fig. 70: Alpha-synuclein

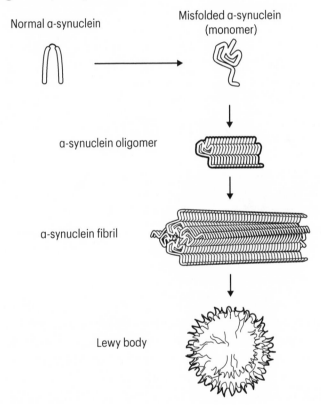

The most sinister aspect of the alpha-synuclein fibrils is that they can spread from one neuron to another. This explains why Parkinson's gradually extends into all parts of the brain.

Fig. 71: Disrupted immune response

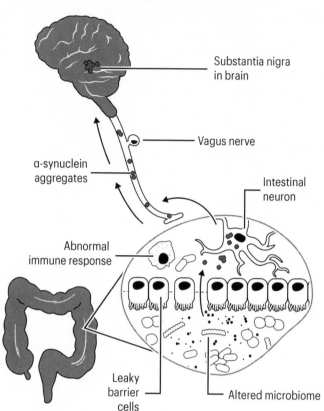

A recent theory gaining momentum is that in many cases, Parkinson's begins not in the brain but the gut. The intestinal tract has many nerve fibers that control the rhythmic contractions that gently move food through the stomach, into the intestines, and then into the colon. As shown in figure 71, it appears that the misfolding of alpha-synuclein can begin in a nerve ending in the gut, perhaps because of the action of bacteria. These nerve endings eventually group together to form the vagus nerve as it travels back to the brain. The fibrils resulting from misfolded alpha-synuclein proteins can travel up this nerve to the brain to the substantia nigra, where they begin to interfere with various aspects of motor function. This phenomenon may help explain the strong association between Parkinson's and autoimmune disorders. One study found that patients suffering from any of a variety of autoimmune diseases had a 33 percent increased risk of developing Parkinson's. This immune response can be triggered by bacteria in the gut, especially if the intestinal lining is disrupted by inflammation.

Yet another path by which Parkinson's may develop involves the mitochondria. Normally, the cell constantly monitors mitochondria and detects when they start to get old and dysfunctional. It does this by tracking the voltage difference (membrane potential) between the inside and the outside of the mitochondria. Remember that the electron transport chain works by pumping protons (H+) from the interior of the mitochondrion into the space between its inner and outer membranes. This makes the inside more negative than the outside. When this difference drops past a certain point, the mitochondrion is tagged for recycling.

Two genes are especially important in this process, and pathological variants are associated with greatly increased risk of Parkinson's. Pink1 is a protein that attaches to the outer membrane of old, damaged mitochondria. When Pink1 builds up on the surface of a mitochondrion, it attracts another protein called Parkin. Have you seen how dead trees are sometimes marked for removal by brightly colored tape? Parkin's role is similar. It attaches a "recycle this one" tag called ubiquitin to the surface of mitochondria that have been pointed out by Pink1. The mitochondrion thus tagged is recycled in a process called mitophagy (a form of the autophagy process that we'll cover later). Pathogenic variants of either the Pink1 or Parkin genes lead to an accumulation of dysfunctional mitochondria. We don't know why this first affects the dopaminergic neurons in one small part of the brain (the substantia nigra), but it might be that those neurons have high energy requirements and are always operating close to their limit. A slight fall in the ATP available to them might cause them to die, resulting in Parkinson's.

There's one other interesting aspect about Parkinson's that illustrates the body's incredible resilience. By the time the symptoms of Parkinson's become evident, more than half of the neurons in the substantia nigra have died. Until this point, that small network of neurons simply rebalances its activity each time it loses a neuron. This compensatory activity is able to stave off symptoms until the substantia nigra reaches a tipping point at which it can no longer continue to operate normally. We see this sort of behavior throughout the body in many different disease states.

Amyotrophic Lateral Sclerosis (ALS)

ALS is another neurodegenerative disease, somewhat similar to Parkinson's. It is often referred to as Lou Gehrig's disease after the famous baseball player who contracted it. Whereas Parkinson's affects relay nerves in the brain, ALS causes the degeneration of the "last mile" neurons used to activate muscles. Muscles are controlled by a type of nerve cell called motor neurons. The vast preponderance of ALS cases are sporadic, but about 5 to 10 percent run within families. ALS is a very serious disease with an average life expectancy of less than five years after diagnosis. Patients quickly lose muscle function all over their bodies, which leads to problems walking, talking, swallowing, and even breathing. The disease tends to strike in the mid-fifties and later, with slightly more men than women being affected.

Like Parkinson's, ALS appears to be caused by the aggregation of misfolded proteins. In at least some cases, this is due to a person having a mutation in the DNA recipe for a gene called SOD1. SOD1 is a critical part of the cell's antioxidant defenses and produces a protein called superoxide dismutase (SOD). As you might recall from previous chapters, superoxide is a free-radical form of oxygen that is capable of damaging proteins, DNA, and lipids within the cell. It is generally produced by mitochondria as a by-product of respiration. SOD is one of a family of enzymes that turn superoxide into the less dangerous hydrogen peroxide that can be dealt with by other enzymes. ALS is especially pernicious because not only do motor neurons build up protein aggregates, but these aggregates rob the motor neurons of their ability to deal with free radicals, which increases the rate of protein damage and misfolding even further.

Fig. 72: ALS

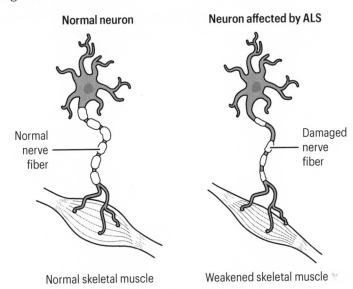

Normal neuron

Neuron affected by ALS

Normal nerve fiber

Damaged nerve fiber

Normal skeletal muscle

Weakened skeletal muscle

Autoimmune Disorders

We saw in the last chapter that the myeloid-lymphoid balance becomes skewed with age. This leads to higher levels of inflammation as the myeloid branch of the immune system comes to predominate. Another, more subtle effect is the disruption of normal balance within the lymphoid arm. One critical class of T cells is called Tregs or regulatory T cells. These lymphoid cells serve to restrain the immune system and help keep it under control. Regulatory T cells play an especially important role in preventing the immune system from attacking the body's own cells.

This whole picture is a setup for autoimmunity. The overall level of inflammation in the body increases with age, putting poorly trained soldiers in all tissues with their fingers on the trigger. If they so much as suspect that one of the body's own cells has a suspicious protein on its surface, they open fire. The regulatory T cells that normally keep the aggressive myeloid cells in check diminish in number as we get older. The chart in figure 73 shows how the presence of antibodies against normal components of the cell nucleus increases with age. These anti-nuclear antibodies (ANA) are frequently used to detect autoimmune diseases like systemic lupus erythematosus (SLE).

Fig. 73: Age graph, antinuclear antibodies

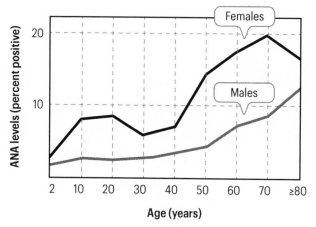

Adapted from Guo (2014).

As we age, some proteins and other cellular components that are ordinarily hidden from the immune system are slowly exposed. For example, wear and tear of joints can expose proteins that the immune system doesn't see when we're younger. As described below, the incidence of misfolded proteins also increases with age. This can result in aggregations of proteins that also look suspicious to the immune system. Combined with the factors mentioned above, the chance of friendly fire increases dramatically.

Osteoporosis

Another common disease of aging, especially among women, is the metabolic bone disease *osteoporosis*. Most people think of bone as inert structural material, not as living tissue. However, bones are extremely dynamic and are constantly being remodeled. This is why they grow and change shape in response to weightlifting and other forms of exercise. Two main types of cells are involved in this process. *Osteoclasts* break down bone tissue while *osteoblasts* create new bone. In midlife, these opposing efforts are finely balanced with little net change. In older adults, however, the rate of breakdown can start to exceed the rate of formation, leading to a gradual loss of bone tissue and a decrease in bone strength.

Fig. 74: Osteoporosis

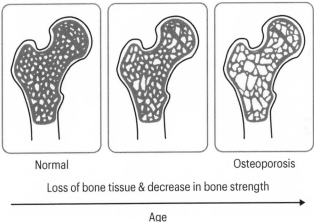

Normal   Osteoporosis

Loss of bone tissue & decrease in bone strength

Age

Until recently, osteoporosis was thought to be driven by an estrogen deficiency in older women, along with poor diet and low levels of vitamin D. Increasingly, it's starting to look like osteoporosis has an important immune component. Inflammation from various causes (autoimmune disease, cancer, etc.) tends to increase breakdown by osteoclasts and decrease new bone formation by osteoblasts. Estrogen probably plays its role in

osteoporosis at least in part by this immune pathway. It is important in the maintenance of regulatory T cells, so a lack of estrogen
leads to a reduction of these peacekeepers. This in turn increases
osteoclastic breakdown of bone. The possibility of transplanting
regulatory T cells into patients suffering from osteoporosis is a
new and exciting direction.

Fig. 75: Osteoporosis flow chart

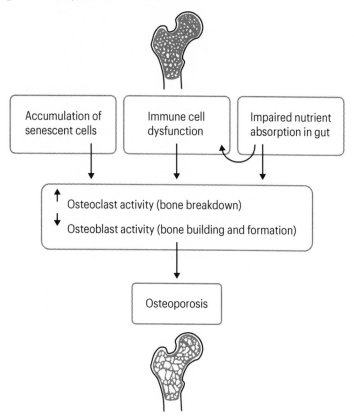

We're also coming to understand the role that gut bacteria (the
microbiota) play in the bone-immune system drama. Microbes
in the gut ferment dietary fiber into short-chain fatty acids like
butyrate, a string of four carbons with a carboxyl (carbon dioxide)

group on one end. These short-chain fatty acids are absorbed into the bloodstream and can have powerful distant effects—including dampening osteoclast activity and thus reducing bone breakdown.

Another contributor to osteoporosis that you'll recognize from the previous chapter is cell senescence. As discussed previously, various stressors can cause cells to shut themselves down to prevent further division. Senescent cells spew a variety of toxic molecules including cytokines into their local environment, increasing inflammation in the area. In bone, senescent cells seem to excite osteoclasts and increase their rate of bone destruction.

Osteoporosis can be difficult to diagnose. One of the primary tools used by physicians is called dual-energy X-ray absorptiometry, abbreviated as DEXA or DXA. This is an X-ray scan that can be used to calculate bone density. These scans are also used for body composition analysis and measuring body fat content. They use very low doses of radiation and are therefore safer than many other imaging techniques.

Sarcopenia

Sarcopenia is a gradual wasting away of muscles, including cardiac muscle in the heart, along with a reduction in muscle function. When someone's cause of death is listed as "old age," sarcopenia is often the underlying explanation. Most people lose anywhere from 3 to 5 percent of their muscle mass every decade once they turn thirty. An eighty-year-old may have lost half of the muscle they had as a young adult.

Muscle isn't just important for looking good in a bathing suit; it is critical to keeping our hearts beating, our lungs breathing, and our bodies mobile. Sarcopenia can manifest in ways other than weakness. For example, people with sarcopenia have more falls and broken bones. Perhaps because the most common cause of sarcopenia is aging itself, it isn't given nearly the attention it deserves. That situation has started to change in recent years with the discovery that resistance training can dramatically improve quality of life for older people, allowing them to get out of the chair and do the things they enjoy. It is often said that humans are similar to sharks in that if they stop moving, they die.

The mechanisms for muscle loss are many and varied. It can be due in part to loss of nerve cells conveying signals from the brain to muscle tissue. If muscle cells don't receive input from the brain telling them to contract, they slowly atrophy. Another cause is lower levels of hormones such as testosterone and growth hormone. These hormones are important indicators to the body that it's time to make more of the proteins needed to form new muscle tissue, as well as to repair existing muscle cells. Sometimes the problem is with protein synthesis itself. Actin and myosin are the main proteins used in the contraction of muscle. If muscle cells don't manufacture sufficient quantities of these and other proteins, no new muscle fibers are constructed to replace those that are inevitably lost over time. The global inflammation that usually accompanies aging is also a prime cause of sarcopenia. And if you need one more reason to avoid obesity, it, too, seems to accelerate sarcopenia as excess fat is deposited in muscle fibers. Adipose (fat) cells are potent producers of inflammatory molecules called cytokines.

Under a microscope, it's easy to recognize sarcopenia in biopsies of muscle tissue. Instead of the dense network of muscle fibers observed in young adults, samples from older people show a dramatic loss of muscle cells and extensive infiltration of fibrous (scar) tissue and fat. Muscle cells require lots of energy to function, and biopsies of muscle tissue from older adults show fewer mitochondria.

Muscle tissue has a special type of stem cell called satellite cells. In younger people, these cells respond to microscopic tears in muscle fibers due to exertion by dividing and specializing to make new muscle cells and fibers. Muscle tissue from older people has fewer satellite cells, especially of the type II or fast-twitch type. This makes it much harder—but not impossible—for older adults to increase their muscle mass.

A good indicator of muscle integrity is grip strength. A simple device called a dynamometer measures how forcefully a patient can squeeze with each hand.

Fig. 76: Sarcopenia chart

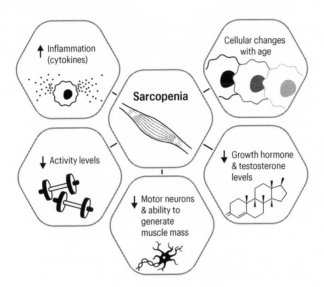

A relatively new way to monitor sarcopenia relies on a blood test for C-terminal agrin fragment (CAF), which results from the breakdown of motor neurons that attach to muscle fibers to tell them when to contract. For reasons that are unclear, CAF levels seem to be a better indicator of sarcopenia for men than for women.

Cancer

Cancer is relatively rare before the age of forty and increases rapidly in incidence thereafter. In the last few decades, we've come to realize that it is really a genetic disease. The DNA in our cells is under constant assault. Breaks in the delicate double helix can be caused by ultraviolet radiation from the sun, chemical pollutants from the food we eat, or reactive oxygen species produced by mitochondria in the normal course of energy production. All of these can lead to mutations that change the sequence of DNA letters in our chromosomes. These mutations are especially likely when cells divide, so we find that the vast preponderance of cancers are in tissues with high rates of cell turnover—skin, intestinal linings, breast, prostate, etc.

Even under normal circumstances, our DNA copying machinery makes an error about one letter out of a billion. This means that a handful of errors can be introduced every time a cell divides. It's only a matter of time until such an error hits the gene for a protein that prevents a cell from dividing unless it has been told to do so, or incapacitates a protein that normally helps repair DNA errors. Such a "hit" can cause one cell to go rogue—to start growing faster than it should. This leads to an army of such cells, called a clone. By dividing more quickly, the cells in that clone now have an increased chance of developing another mutation that takes off another brake. So now there's a subclone that is growing even faster than before. After a few such mutations, the result is a line of cells that have lost all normal growth controls—a cancer.

This increased probability of incipient cancers must be considered against the backdrop of a declining immune system that is less able to detect and put down the rebellion. As if that weren't enough, we've seen how age tends to increase insulin signaling, which provides a powerful push to cancer cells. All the many factors associated with aging—inflammation, transcriptional noise, shorter telomeres, and others—create a perfect storm that results in a constant torrent of rogue cells. Soon or later, cancer ensues.

IT MAY be depressing or even frightening to consider the various diseases that tend to develop as we age. But as we'll see in the next section, we don't have to simply accept this situation. LUCA, our ultimate ancestor, bequeathed us a powerful program hidden in plain sight in our DNA that can help us stave off the ravages of aging. Let's learn about it now.

References and Further Reading

U Foger-Samwald et al (2020). Osteoporosis: pathophysiology and therapeutic options. *EXCLI Journal*, 19: 1017–37.

GR Geovanini and P Libby (2018). Atherosclerosis and inflammation: overview and updates. *Clinical Science*, 132: 1243–52.

YP Guo (2014). The prevalence of antinuclear antibodies in the general population of China: a cross-sectional study. *Current Therapeutic Research*, 76: 116–19.

R Henry (1998). Type 2 diabetes care: the role of insulin-sensitizing agents and practical implications for cardiovascular disease prevention. *American Journal of Medicine*, 105(1A): 20S–26S.

MJ Rae et al (2010). The demographic and biomedical case for late-life interventions in aging. *Science Translational Medicine*, 2(40): 40cm21.

C Raza et al (2019). Parkinson's disease: mechanisms, translational models and management strategies. *Life Sciences*, 226: 77–90.

CD Rietdijk et al (2017). Exploring Braak's Hypothesis of Parkinson's disease. *Frontiers in Neurology*, 13 February.

RA Sperling et al (2011). Toward defining the preclinical stages of Alzheimer's disease: recommendations from the National Institute on Aging–Alzheimer's Association workgroups on diagnostic guidelines for Alzheimer's disease. *Alzheimer's & Dementia: The Journal of the Alzheimer's Association*, 7(3): 280-292.

E Tan, Y Chao, A West et al (2020). Parkinson disease and the immune system: associations, mechanisms and therapeutics. *Nature Reviews Neurology*, 16: 303–18.

JD Walston (2012). Sarcopenia in older adults. *Current Opinions in Rheumatology*, 24(6): 623–27.

PART THREE

An Ancient Program Hidden in our Genes

9

Cell Signaling

"Some cultural phenomena bear a striking resemblance
to the cells of cell biology, actively preserving themselves
in their social environments, finding the nutrients they
need and fending off the causes of their dissolution."

DANIEL DENNETT

IN THIS chapter, we'll examine the internal life of the cell and even eavesdrop a bit to see how the various components talk to each other. It's impossible to appreciate how our bodies respond to various types of stress without understanding what's going on within the cell. Always keep in mind that each of the cells in our bodies is a descendant of fiercely independent organisms that previously existed on their own. There's a constant tension between the needs of our individual cells and those of our body as a whole.

Recall that each cell is encased in a fatty bubble—the cell membrane. Entry and exit through that membrane is tightly controlled. Smaller hydrophobic molecules and gases like carbon dioxide can pass right through. However, larger molecules and those carrying a positive or negative charge need some sort of tunnel through the membrane to get into the cell. This is why cells need to insert a special door in their membranes to allow glucose to enter.

From the twenty-thousand-foot level, cells are pretty dumb. Think of them as tiny computers constantly executing the following program:

```
while (true) do:
    if (ATP is sufficient) AND (glucose is sufficient) AND (protein is
    sufficient) then
            sufficient_resources = true
    end
    if GrowthFactorsDetected() AND sufficient_resources then
            Grow()
            Divide()
    end
end
```

But how do they know if resources like ATP and glucose are sufficient? How do they detect growth factors?

The cell membrane is studded with proteins. Some serve as channels to allow molecules like glucose in and others like lactate out. Other proteins act as antennae, scanning the outside environment for particular molecules. We call these proteins *receptors* because their job is to detect a signal such as insulin on the outside of the cell and then initiate some response on the inside. Protein receptors can do this because they have one end sticking out into the external environment and another poking into the cytoplasm of the cell.

The circles and triangles in figure 77 represent molecules outside the cell. When one of these molecules binds to a receptor, it can cause a change inside the cell—the release of the oval. In some cases, this generates a molecule that can travel to the nucleus to cause certain genes to be turned on or off.

In addition to receptors that are scanning the outside world and alerting the interior of the cell when they come across something interesting, there are other proteins that are monitoring conditions inside the cell. The main ones we're going to talk about are AMPK and mTOR.

Fig. 77: Protein receptors

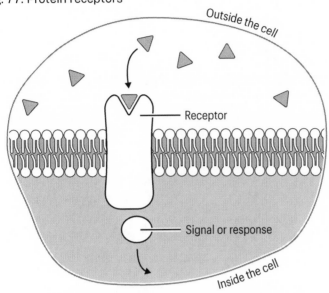

Fig. 77: Protein receptors

The first thing the cell needs is an energy sensor, kind of a fuel gauge to tell it whether it's running low or not. This role is mainly filled by a protein called AMPK. You can think of the primary fuel molecule in the cell, ATP, as a tiny battery that can be charged up, used to power an operation like building a protein, and then recharged again when depleted. AMPK is constantly monitoring the ATP level in a cell. When ATP runs low, AMPK sounds an alarm by putting a tag on other proteins that can turn them either on or off.

While AMPK is monitoring the cell's energy levels, mTOR is scanning for the availability of building blocks, especially amino acids. In times of plenty, mTOR sees lots of amino acids floating around and tells the cell, "Let's grow!" Like AMPK, it does this by tagging other proteins in a relay that ultimately makes its way to the nucleus of the cell where the DNA cookbook is stored. mTOR thus activates a complicated program, turning some genes on and others off in a way that causes the cell to grow like mad and get ready to divide.

These two sensors, AMPK and mTOR, interact in an important way. AMPK turns off mTOR. If you think about it, this makes total sense. It takes both energy and amino acids to build a new protein. If you have amino acids, but not enough energy, you won't be able to build anything. So when AMPK is activated in times of energy scarcity, mTOR is turned off regardless of whether amino acid building blocks are available or not. Since energy in cells mainly starts with the burning of glucose (sugar), AMPK activation indirectly reflects the amount of glucose around. When both glucose and amino acids are low, AMPK is active and mTOR is completely shut down.

Fig. 78: AMPK

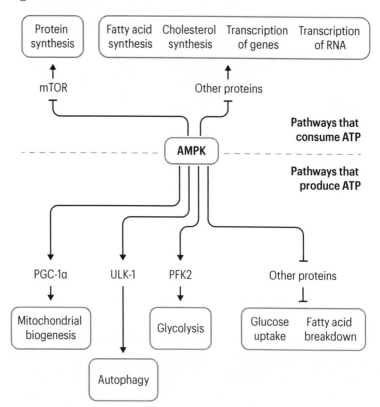

We can see from figure 78 that AMPK is a critical power broker in the cell. When it's activated (in a low-energy environment), the cell gets busy burning whatever it can to make energy. AMPK increases insulin sensitivity by signaling for more GLUT4 glucose transporters to be inserted into the cell membrane. AMPK also prompts the production of more mitochondria and an increase in a type of recycling activity called autophagy that we'll cover later. AMPK is just as important for the activities it turns off—laying down new fat and other construction projects. The activation of AMPK is a signal to the cell that there might be tough times ahead.

Fig. 79: Autophagy

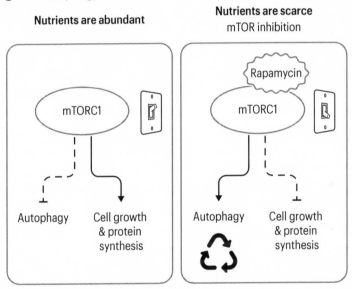

As mentioned, mTOR is the other critical nutrient sensor, primarily for the cell's amino acid supply. mTOR is the main component of a multi-protein complex called mTORC1. If there are sufficient building blocks available and enough energy, mTOR signals the cell to synthesize new proteins and grow. It also turns off autophagy. When mTOR is inactivated because nutrients are in scarce supply, autophagy is turned back on and protein synthesis is disabled.

The ability to turn on AMPK and turn off mTOR seems to be at the heart of many of the pathways that respond to mild stressors such as fasting and exercise. These stressors initiate multiple adaptive programs, including autophagy (recycling), mitochondrial biogenesis (production of new mitochondria), and production of antioxidants and molecular chaperones.

The Sirtuins

Another set of proteins are critical in the cellular response to stress—the sirtuins. These proteins were discovered during early investigations of the mechanism of caloric restriction in yeast. Mammals have seven similar sirtuin proteins called SIRT1 through SIRT7. The SIR moniker derives from the phrase *silent information regulator* because it was found that these proteins help cells keep their DNA properly wrapped around the histone spools.

It's important for cells to keep some genes turned on and others turned off, depending upon the function of the cell. As we age, the DNA machinery gets "leaky." Some genes that should be kept wrapped tightly around their spools and completely turned off start to loosen and produce low levels of protein. For a physical metaphor, imagine a brand-new refrigerator that has tight door seals and maintains its temperature within a narrow range. As that refrigerator gets older, the door seals begin to leak and the temperature inside the refrigerator varies. Food kept in the refrigerator doesn't last as long and isn't kept as fresh. This is what seems to happen in our cells. In youth, the difference between a liver cell and a heart cell is stark. Certain proteins are completely turned on or off in each type of cell. As we age, however, this sharp picture starts to blur. There are increasing amounts of transcriptional "noise"—proteins that shouldn't appear at all within a given type of cell now start to be produced at low levels.

While many genes are supposed to be permanently disabled as cells differentiate, others need to be turned off only temporarily, until they are needed again. One important method of controlling the degree of compactness of a region of DNA is by attaching or removing two-carbon acetyl groups on the spools. When acetyl groups are attached to a spool, the DNA looped around it opens up and is available for reading. When the protein coded by that DNA is no longer needed, the acetyl groups can be removed from the associated spool, causing the DNA to once again wrap up tightly.

The sirtuins fall into the category of histone deacetylases. That is, they remove the acetyl groups from the histone proteins that comprise the spools and thus silence the associated DNA. Where do the two-carbon acetyl groups come from? Food! When food is plentiful, there are more acetyl groups available to loosen DNA. In fact, one of the most important sirtuins, SIRT1, is activated by AMPK. Recall that AMPK is itself activated by a low-energy state. So, under fasting conditions, SIRT1 does a better job of shutting down genes that should be shut down.

Another tie to metabolism comes from the fact that the sirtuins require NAD+ to operate. In fact, they work by transferring an acetyl group from a protein (such as a histone spool) to NAD+. So each operation performed by a sirtuin uses up one molecule of NAD+. Under low-energy conditions, NAD+ levels rise within a cell. So once again, fasting makes it easier for the sirtuins to do their critical jobs.

Unfortunately, NAD+ levels decrease with age. This means the sirtuins are less effective as we get older. Even worse, they have to compete with another set of proteins that also require NAD+ to operate. The PARP proteins are DNA repair enzymes. Because DNA breaks become more common as we age, the PARPs compete more and more with the sirtuins for the diminishing pool of NAD+. For these reasons, it becomes even more important to periodically boost the sirtuins via caloric restriction with the passage of decades. As covered in the upcoming chapter on recommendations, older adults might also consider taking supplements proven to boost NAD+ supplies.

Fig. 80: Sirtuins & stress

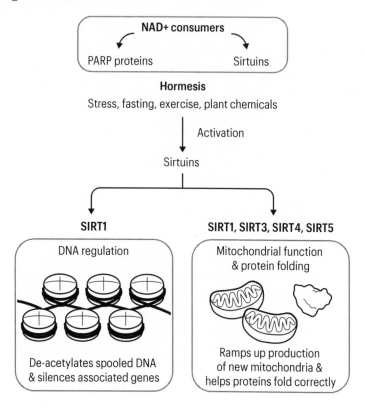

The impact of the sirtuins goes beyond keeping certain genes turned off. Three of the sirtuins—SIRT3, SIRT4, and SIRT5— operate inside mitochondria, where they help optimize mitochondrial function. SIRT1 is a critical stress-response protein in the rest of the cell. In fact, AMPK activates SIRT1 to help communicate the "low energy" alert throughout the cell. SIRT1 in turn helps activate the recycling process known as autophagy. SIRT1 also helps ramp up the production of new mitochondria along with antioxidants and the chaperones that assist new proteins in folding correctly. It is becoming evident that all roads involving hormesis (a further exploration of which is upcoming) lead to SIRT1. In other words, intermittent fasting, exercise, plant chemicals, and many other mild stressors operate at least in part by activating SIRT1.

NRF2

. . . .

Another well-known stress response pathway involves the proteins KEAP1 and NRF2. NRF2 is a transcription factor—a protein that can travel to the nucleus and bind to certain gene promoters to initiate the production of various other proteins. As shown in the following diagram, under normal conditions, NRF2 is kept in the cytoplasm by a pair of KEAP1 proteins. These proteins bind to NRF2 and cause it to be degraded in a cellular shredder known as the proteasome. However, when oxidative stress becomes high enough, the sulfhydryl groups (-SH) on the KEAP1 proteins are oxidized, causing the two proteins to link together via a disulfide (S—S) bridge. This linkage disconnects the KEAP1 proteins from NRF2, freeing it to travel into the nucleus and turn on a number of genes containing the ARE (antioxidant response element) sequence. NRF2 can also be activated by ultraviolet radiation, toxins, pathogens, inflammation, and other stressors.

Fig. 81: NRF2

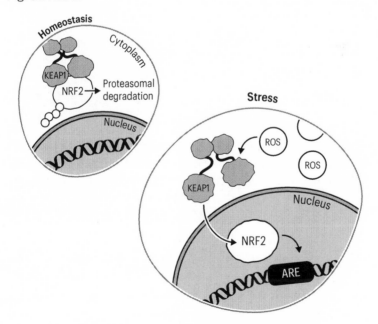

No matter how it's activated, when NRF2 enters the nucleus, it triggers the production of a variety of protective proteins. Some detoxify reactive oxygen species (ROS) while others act as chaperones to help misfolded proteins arrive at the desired shape. As with the sirtuins and other stress response players, NRF2 can have both positive and negative impacts in cancer. NRF2 can prevent cancer by helping normal cells avoid the mutations caused by oxidative damage. However, cancer cells can co-opt NRF2 to assist in their own survival. The simplistic view is that NRF2 is helpful before the development of cancer but may be harmful once cancer has taken root.

We'll talk about the varied roles of plant chemicals in a later chapter, but it's interesting to note that cruciferous vegetables like broccoli contain chemicals called sulforaphanes that can activate NRF2. This may help explain some of the health benefits of these plants. They mimic oxidative stress to a small degree and prompt the cell to take preemptive measures to be ready for larger stressors that might be coming. A recurring theme that you may note in this book is that by preparing for future disasters, cells might become more resistant to a wide variety of stressors.

Insulin Signaling

Insulin is a small protein hormone that the beta cells of the pancreas secrete into the blood when they sense a high enough level of glucose. Most cells in the body contain receptors in their cell membranes that monitor the environment for insulin. When insulin binds to two of these receptors, they move together and trigger a signaling cascade inside the cell.

Note that the binding of insulin to the receptor kicks off two different signaling relays. One acts as a *mitogenic* signal that tells the cell that conditions are favorable to divide. The other works through mTOR to help the cell absorb the glucose now available in the blood and use it to produce energy.

Fig. 82: Insulin signaling

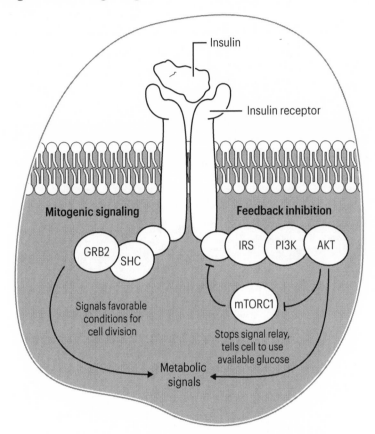

As we've seen previously, mTOR is a key sensor of the nutritional status of the cell. When it is activated, the cell ramps up protein production and prepares to grow. When mTOR is deactivated, the cell turns off protein synthesis and steps up its internal recycling.

One important effect of insulin binding to its receptor is that it causes the cell to insert GLUT4 glucose transporters into its membrane so it can let the available glucose in.

Insulin resistance results from a failure of insulin binding to trigger opening of the GLUT4 doors.

Fig. 83: Insulin resistance

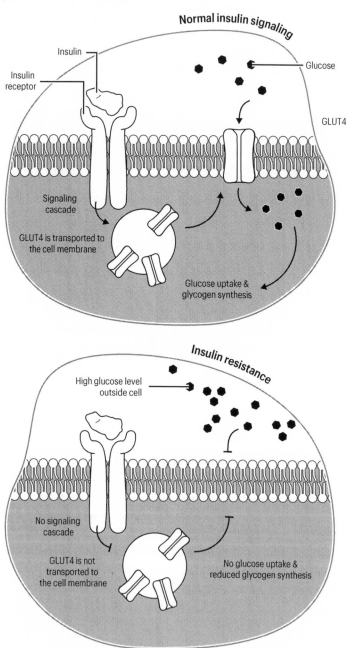

Cell Signaling and Longevity

Just about every anti-aging intervention discovered so far works by affecting cell signaling. The signal transduction pathways within cells are extremely complicated, but we can oversimplify with this summary of possible ways to reduce the rate of aging.

Objective	Effect	Interventions
Activate AMPK	When activated, AMPK activates SIRT1 and SIRT3, which improves mitochondrial function and causes the cell to get ready for more burning of fat to produce energy.	Caloric restriction and exercise. The anti-aging and anti-diabetes drug metformin also activates AMPK.
Inhibit mTOR	mTOR increases protein synthesis and other growth-oriented activities within the cell. These measures are helpful in youth but work against us in age. mTOR inhibits the recycling process of autophagy.	The anti-aging drug rapamycin inhibits mTOR. AMPK also inhibits mTOR, so mTOR is indirectly reduced by caloric restriction and exercise.
Activate SIRT1/ SIRT3	SIRT1 and SIRT3 activate various stress-defense systems within the cell and increase the number and quality of mitochondria in preparation for increased fat burning.	Polyphenols such as resveratrol can activate SIRT1 and SIRT3. These proteins require NAD+, so NAD precursors such as NR and NMN help increase sirtuin activity.
Activate NRF2	NRF2 is activated in times of oxidative stress. It increases the production of chaperones, antioxidants, and other defenses against oxidation.	NRF2 is activated by the sulforaphanes in broccoli sprouts and other plants.

Objective	Effect	Interventions
Activate FOXO	FOXO increases the rate of autophagy and thus the cleanup of misfolded proteins.	The FOXO proteins are activated by AMPK and SIRT1.

Summary

Given the increasing problems with misfolded proteins and their aggregates that accompany age, increasing autophagy is under intense investigation as a means of improving health and lengthening lifespan. Autophagy is a common downstream effect of many of the hormetic stress responses described in this book. It helps ameliorate several of the hallmarks of aging, including oxidative stress, proteotoxic stress, and decreased mitochondrial function. Several methods for increasing autophagy have been shown to result in improved health and longer lifespans in laboratory experiments. More on that topic soon.

References and Further Reading

L Bosch-Presequé and A Vaquero (2013). Sirtuins in stress response: guardians of the genome. *Oncogene* 33, 3764–75.

P Hiebert and S Werner (2019). Regulation of wound healing by the NRF2 transcription factor—more than cytoprotection. *International Journal of Molecular Sciences*, 20(16): 3856.

10

A Secret Program

THE WORD that every biology student learns regarding life is *homeostasis*. This term describes the tendency of most biological parameters to operate within a narrow range and the ability of organisms to take corrective actions when parameters start to fall outside of that range. For example, if your blood pressure drops when you stand up, your body will constrict arteries and your heart will beat harder to restore a normal blood pressure.

There's no dispute that homeostasis is a critical feature of life. Our bodies are constantly working to keep all operations within some normal range. One aspect of resilience is the ability to withstand temporary excursions of various parameters outside their normal ranges in response to a stress such as an infection and to bring them back in bounds. However, homeostasis is usually used in a fairly narrow sense—at the level of individual parameters (like temperature) or specific aspects of physiology (e.g., fluid balance). Another term has arisen to describe the overall balance of all systems within the body—*allostasis*.

Let's go back to a physical metaphor we used in an earlier chapter—the body as a trampoline. Homeostasis is the tendency of the trampoline to push back against a load placed upon it so it returns to its original position. But what if the load isn't temporary?

Allostasis

Allostasis is the concept of homeostasis applied across all the systems in an organism. It takes into account the fact that long-term changes in one system can cause lasting changes in other systems. We see this in our bodies when long-term high blood pressure causes the heart wall to thicken in order to pump against the greater pressure. We can also think of allostasis as kind of a "new normal." In our trampoline example, imagine a heavy weight placed in the middle of the trampoline bed and left there for several weeks. After a while, the new normal for the trampoline is to have a saggy bed where the weight sits. Even after we remove the weight, the trampoline bed no longer returns to its original position. In the same way, if we constantly overeat, we eventually reach a new normal that involves more fat stored in our bodies, a higher insulin level, and many other unfavorable changes. Reversing this state requires work to achieve a different new normal.

Hormesis

While it has taken the biomedical community a while to accept the notion of allostasis, the concept of *hormesis* has had an even rockier road. The term was first coined by researchers back in the 1940s to describe a strange phenomenon. Toxicologists studying various poisons tried to graph the negative effects of those poisons at different doses. They expected a linear dose response looking something like this:

Fig. 84: Hormesis/dose response curves

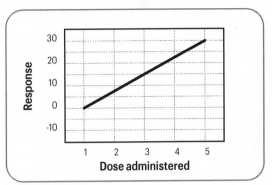

The meaning of this expected curve is that the higher the dose of poison, the greater its adverse effect (response). Instead, they observed curves like this:

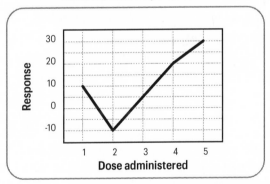

This is called a U-shaped dose response curve, and it suggests that there's a dose range that is actually beneficial. As you can imagine, this result didn't sit well with people trying to get toxic chemicals banned. Could the petrochemical industry use it to argue that pollutants can actually be good? These findings were extremely controversial, and many scientists assumed that they were measurement errors.

However, the same U-shaped dose response curve kept being observed in wildly different contexts: fruit flies exposed to high or low temperatures, worms subjected to varying levels of radiation, or even mice deprived of oxygen in varying degrees. As hard as it was to explain, a wide variety of agents that should be purely damaging seemed to have a range in which they evoked beneficial responses within many different organisms. What in the world could be going on?

The idea behind the phenomenon known as hormesis is relatively simple. When a cell or organism is exposed to low levels of a stressor (chemicals, radiation, etc.) for relatively short periods of time, it takes steps to adapt to that stressor. What's more, the corrective steps taken are often in excess of the degree of the stressor. It's as if the cell or organism is getting ready for even worse to come. From an evolutionary standpoint, this makes total sense. If an animal starves for a few days, it prepares for an even longer period without food. If it encounters low temperatures, it gets ready for a prolonged period of cold. When we lift weights, our muscles grow bigger and stronger in the expectation that they'll be asked to do even more.

In future chapters, we'll talk about hormesis at the level of the organism. But to understand what's going on, we have to first dive down to the level of the cell. Let's look at the various stressors that cells see and what measures they take to counter them.

Hypoxic Stress

Most eukaryotic cells, like those that comprise our bodies, are extremely sensitive to the amount of oxygen available to them. They have no means of storing oxygen for a rainy (or anoxic) day and are reliant on a constant fresh supply. Recall that transcription factors are proteins that are able to enter the nucleus and turn on or off various genes by attaching to their promoters—sections of DNA that the transcription machinery binds to when it's time to make an mRNA copy of the gene that can be used as a protein recipe by the incredible ribosome protein factories. We'll see several examples in this book in which a transcription factor also acts as a sensor. This is an efficient architecture because the same sensor can also cause changes within the cell by affecting DNA transcription. It's kind of like a motion sensor that can turn on an alarm siren.

Hypoxia inducing factor one alpha (HIF1a) is one such sensor. Normally, this protein sits in the cytoplasm and is tagged for destruction almost as fast as it's made. The trick is that the tagging happens only in the presence of sufficient oxygen. If oxygen levels in the cell drop past a certain point, HIF1a can't be destroyed and starts to build up in the cell. Quite soon, some HIF1a molecules start to make their way into the nucleus, where they bind to promoters called HIF response elements, or HREs. For readers with programming experience, it's as if we have several blocks of code surrounded by the condition "If HIF1a > 0 Then..." The buildup of HIF1a caused by hypoxia (a lack of oxygen) causes this condition to be true. The code blocks represent the genes controlled by the HRE promoters.

The genes activated by HIF1a have a variety of effects throughout the cell and the entire body. These include increasing glycolysis, a way of producing ATP from sugar without using oxygen. At the level of the body, HIF1a activation leads to the production of the hormone erythropoietin (EPO), which increases the production of red blood cells by the bone marrow. The growth factor VEGF is also produced, which causes the growth of new blood vessels in an attempt to provide more oxygen.

Exercise is one behavior that causes mild hypoxic stress. During intense activity, muscles quickly run out of oxygen and activate HIF1a. High altitude is another trigger for HIF1a. Regardless of how it's activated, HIF1a causes an increase in the expression of various genes whose effects can last for hours or days.

Oxidative Stress

By now, oxygen's two-faced nature should be clear. When properly harnessed by mitochondria, oxygen acts as the final electron receptor and drives the production of energy in the form of ATP. Unfortunately, even billions of years of evolution don't lead to perfection, and oxygen occasionally wreaks havoc within a cell. We've already learned what a villain oxygen can be: when it fails to sit patiently at the end of the electron transport chain where it can be carefully loaded with electrons, its rogue behavior leads to the creation of molecules called reactive oxygen species (ROS)—or, more colorfully, free radicals.

In chemistry, a radical is an atom or molecule that has a highly reactive unpaired electron. Such radicals can attach themselves to almost anything in the cell—the membrane, proteins, even DNA. Over the eons, our cells have evolved powerful machinery to mop up these oxygen radicals before they can do damage. Two such mop-up crews are superoxide dismutase (SOD) and catalase. However, eukaryotic cells have learned to recognize small amounts of ROS as signs that the mitochondria are hard at work. It's a bit like monitoring a factory by looking at the amount of smoke it pushes out its smokestack. If there's too much, we send in the firefighters along with chemical containment crews. If there's just a little, we know everything is operating normally. Now, imagine that there's a huge fire in the factory. What if someone had installed a device a few days ago that could absorb smoke just before it entered the factory's chimney? The fire would burn out of control before anyone was even aware. By the time the evidence showed up in the form of smoke, there would be no saving the factory.

The situation in the cell is similar, and perhaps explains the puzzling lack of benefit provided by taking high doses of antioxidants. When the cell sees ROS start to increase because of exercise or other mitochondrial stressors, it produces special proteins to help deal with the underlying problem. The cell's response might even be to tear down old mitochondria that are spewing out large amounts of pollution. By taking large doses of antioxidants, we interrupt this delicate dance. The result may even be an accumulation of low-quality mitochondria that accelerates the effects of aging. Although we don't understand exactly why, the antioxidants provided by most plants don't seem to have the same adverse effect. It might be because plants produce a wide variety of antioxidants at much lower levels than the megadoses of individual antioxidants that we normally find in a pill.

Fig. 85: Oxidative stress

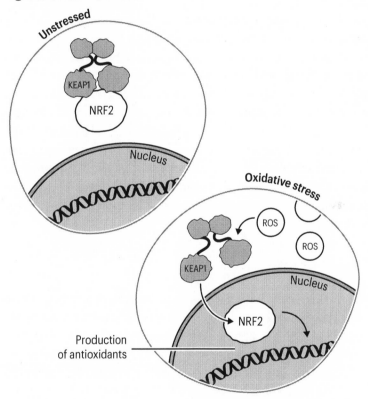

We met one of the key sensor-effector systems for oxidative stress in the chapter on cell signaling: NRF2/KEAP1. As you'll recall, NRF2 is a powerful transcription factor capable of activating dozens, perhaps hundreds, of genes in the nucleus. Normally, however, NRF2 is kept from entering the nucleus by KEAP1. Under oxidizing conditions—when the cell is undergoing oxidative stress—KEAP1 is oxidized and loosens its grip on NRF2. This allows NRF2 to enter the nucleus and kick off a response to the oxidative stress. Part of this response involves the production of various antioxidants such as thioredoxin (breaks down hydrogen peroxide), superoxide dismutase (SOD; turns superoxide radicals into hydrogen peroxide), and heme oxygenase-1 or HO-1 (reduces inflammation in blood vessels).

Another important player for dealing with oxidative stress is the small molecule glutathione. Many proteins are partly held in the right shape by disulfide "bridges." These are reversible S—S (sulfur-sulfur) bonds between two cysteine amino acids within the protein. Under oxidative stress, these bonds can be broken, leading to the protein losing its shape. Glutathione is one of the most important antioxidants made by the cell. It is able to reverse the oxidation of these disulfide bridges and help restore the proper shape to misfolded proteins. Deficits in glutathione function have been observed in people with neurological problems such as schizophrenia and neurodegenerative diseases like Alzheimer's.

Thermal Stress

Mammalian cells have evolved to operate in a very narrow temperature range. The thousands of reactions taking place every second have been optimized to conditions around 36 to 37°C (97 to 99°F). Any significant deviation from this temperature acts as a stress on the cell and elicits the appropriate compensatory measure. The best known is called the heat shock response (HSR). This response can easily be demonstrated in every organism from bacteria to humans. Excess heat is dangerous for life because it interferes with the proper folding of proteins. Cells are extremely dependent on proteins being able to find their proper shapes. Proteins are tiny machines. Imagine a construction crane bent into a V shape by a violent storm. It can no longer do its job, so work at the construction site halts. A cell is jam-packed with millions of proteins at any given time. No wonder life has evolved such an elaborate method of dealing with excess heat.

Mammalian cells make a number of proteins called heat shock factors or HSFs. The best understood is HSF1, which operates throughout the cell. Heat shock factors act as sentinels. If they detect heat stress, they are supposed to run to the nucleus and activate a number of genes that have sequences called heat shock elements (HSEs) in their promoters. This turns on the production of chaperones called heat shock proteins (HSPs). HSP70 and HSP90 are responsible for finding proteins starting to misfold because of the heat stress and helping them maintain their proper shapes. In the absence of heat stress, HSP70 and HSP90 bind to HSF1 and keep it from going into the nucleus. If these chaperones are called away to deal with misfolded proteins, HSF1 is suddenly free to travel into the nucleus, leading to the production of more HSP70, HSP90, and other proteins.

There is actually a whole family of heat shock proteins that work together to deal with misfolded proteins. Some are called holdases and have the job of finding misfolded proteins and simply holding them for another member of the family—a *foldase*. The foldases require ATP and act as chaperones to help the misfolded protein assume its correct shape. Another member of the family is called CHIP. Its job is to identify misfolded proteins that are too far gone and to take them to the shredder, an organelle called the proteasome.

It's important to view the heat shock response as part of a continuum. Proteins are constantly being made. Some fold correctly and go about their work, some don't. During times of stress, protein misfolding increases and the heat shock proteins are sent out to try to help. If some proteins are such tangled messes that they can't be folded correctly, they are tagged for destruction. During times of protein-folding stress, the overall process of recycling (autophagy) is ramped up. We'll have a whole chapter dedicated to autophagy, but think of it as a way of scooping up a bunch of misfolded proteins at the same time, putting them in a garbage bag, and then delivering that bag to an incinerator.

We've focused on heat stress in this section, but cold stress is just as real. It won't surprise you to learn that cells have cold shock proteins as well. We'll talk more about cold stress and the hormetic response to it later.

Protein-Folding Stress

We learned in the chapter on aging that many of the most pernicious chronic diseases feature aggregations of misfolded proteins. In Alzheimer's, clumps of beta amyloid form outside of neurons in the space between them. In Parkinson's, tangles of a different protein called alpha-synuclein accumulate inside neurons and eventually lead to their death.

It turns out that proteins misfold all the time. In youth, we're able to keep the problem under control by producing large numbers of chaperones that help other proteins fold correctly—or escort them to the protein garbage disposal (the proteasome) if they can't be properly folded. The continuous cellular recycling process called autophagy is also more active in youth, keeping the burden of misfolded proteins low.

As we age, the quantity of misfolded proteins and their aggregates begins to rise exponentially. Unfortunately, just when we need our protein maintenance machinery to be on its game, it, too, starts to fail us. We produce fewer chaperones as we age and also do a poorer job of breaking down misfolded proteins and their aggregates. This leads to a condition called *proteostatic collapse.*

The cell senses the stress imposed by misfolded proteins. As with all the stressors we've covered, the body is able to mount a response—up to a point. In healthy younger cells, the detection of unfolded proteins leads to a number of measures to counteract the stress.

Fig. 86: Misfolded proteins with age

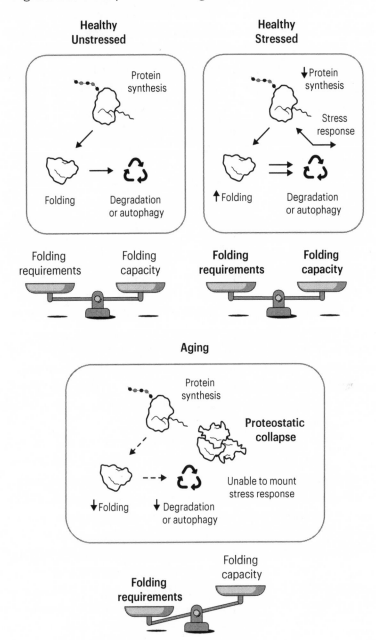

We've already seen some of the ways cells detect protein mis-folding stress because they are the same mechanisms involved in oxidative and heat stress. Both the NRF2/KEAP1 systems as well as HSF1 in combination with HSP70/HSP90 operate the same way. They consist of a transcription factor (either NRF2 or HSF1) that is normally kept in the cytoplasm bound to another protein (KEAP1 or one of the HSPs). Under times of stress, the transcription factor is freed from its keeper and allowed to enter the nucleus where it activates a number of genes that mount the appropriate response. For many types of stress, the response is similar—increase production of chaperones to help proteins fold correctly and augment the rate of autophagy to recycle old, mis-folded proteins.

DNA Damage Stress

Every time a cell divides, it must make a complete copy of its DNA. This involves gently separating the two strands of the double helix and reading along each to make an exact copy. Every cell contains approximately three billion DNA letters (actually times two because there are two chromosomes, one from each parent) that must be faithfully replicated. Mammalian DNA replication is amazingly accurate with fewer mistakes than one in a billion. But a handful of errors are still introduced with each cell division. Most of these errors are harmless since more than half of the genome consists of repeating sequences left over from ancient viruses, but every once in a while, a vital gene is altered. Occasionally this results in a mutation that allows the daughter cells to grow more quickly and can ultimately result in cancer.

However, DNA is under assault even in non-dividing cells. Radiation, environmental toxins, pollution, and just normal life lead to breaks in one or both of the double helices in a cell every minute. Thankfully, we have an elaborate set of repair enzymes that can usually make a perfect correction. As with responding to any problem, the first step in fixing DNA breaks is detecting them. Dozens of genes make proteins that can sense various types of problems in DNA and initiate a fix. Examples include ATM/ATR, MSH2-6, and the MRN complex. An important player in the DNA damage response is a protein named p53 that is often called the guardian of the genome. In dividing cells, p53 doesn't just cause the repair machinery to spring into action—it also delays the cell cycle to give the cell time to fix the problem before dividing. If it senses that things are too messed up, p53 can actually force the cell to self-destruct, a process called *apoptosis*. Cancer cells almost always have to figure out a clever way to disable p53 in order to grow.

The detection of DNA damage sets off alarm bells throughout the cell. In addition to stopping the cell cycle, cells also start to produce more antioxidants, chaperones, and other protective proteins. It may be a head-scratcher, but even mild DNA damage can evoke a beneficial hormetic response. NASA is working hard to figure out how to escalate this response before sending astronauts to Mars, a journey that will expose them to high levels of radiation and intense DNA damage. The preparation for such missions will almost surely involve exposing the astronauts to low levels of radiation on earth to activate hormesis and increase their ability to withstand the DNA damage stress ahead.

Inflammatory Stress

Inflammation is an important form of stress for both individual cells and the body as a whole. We've talked about the potential dangers of oxidation, so it should come as no surprise that our immune systems use oxygen as a weapon against bacteria and other invaders—or that these pathogens also use oxygen against us! During periods of inflammation, cells are thus exposed to increased oxidative stress whether by friend or foe. As with other hormetic stressors, inflammation can actually be beneficial in small, intermittent amounts, like when it accompanies exercise. But chronic or more pronounced inflammation quickly becomes detrimental.

Nutritional Stress

Nutritional or energy stress is perhaps the best-known driver of hormesis. Cells are exquisitely sensitive to the amount and types of nutrition available to them. This topic will be covered extensively in upcoming chapters, including one on intermittent fasting, so we'll just give an overview here.

Cells have multiple systems for detecting levels of nutrition and energy. At the most basic level, a cell must constantly be asking itself, "Do I have enough energy and building blocks for my immediate needs, or should I prepare for a prolonged period of scarcity?" Life is an ongoing battle against entropy—the tendency for things to break down and decay. This battle requires a constant supply of energy. In fact, we can think of life as a process that uses energy to build and maintain complexity. Every cell must continuously build and maintain complex components such as DNA molecules, proteins, and the lipids in membranes. If this process halts for more than a few seconds, the cell's integrity is in jeopardy.

We'll focus here on the cell's biggest ongoing maintenance job: proteins. Proteins do all the hard work within the cell. They act as enzymes to speed up reactions, function as tiny machines to create new components, and serve as scaffolds to maintain the cell's shape. It takes significant energy to build new proteins, help them fold, keep them in the proper shape, and then break them down when they're no longer needed. Although these processes require many ingredients, the two most important are energy (ATP) and building blocks (amino acids). To oversimplify just a bit, the cell uses ATP to weld amino acids together into the long chains we call proteins. So the cell must constantly monitor its supplies of ATP and amino acids. When either of these commodities starts to run low, the cell must take countermeasures that include:

1   Improving the function of existing mitochondria.

2   Producing new mitochondria.

3   Synthesizing more antioxidants and components to deal with increased activity in mitochondria.

4   Slowing down the construction of new proteins.

5   Ramping up recycling activity, which increases the amino acid supply by breaking down old or misfolded proteins.

In other words, nutritional stress causes the cell to change its focus from growth and division to efficiency and survival. This change in focus is decidedly beneficial for us as we age. In particular, we've talked repeatedly about the tendency for older cells to accumulate quantities of misfolded proteins that can eventually create dangerous aggregates inside or around the cell. Aging also leads to an accumulation of raggedy mitochondria that leak reactive oxygen species (ROS). Mild nutritional stress helps counteract these harmful results by initiating the replacement of poorly functioning mitochondria and ramping up the recycling process known as autophagy. Part of the benefit ascribed to exercise and certain plant molecules is due to their ability to mimic nutritional stress. Exercise does this by reducing the amount of ATP within cells (by using it up). Phytonutrients (plant chemicals) act as mild mitochondrial poisons, causing mitochondria to leak superoxide, hydrogen peroxide, and other free radicals. In both cases, the cell reacts to the stress in the same way as it does short-term starvation.

Summary

The trillions of cells within our bodies don't know that it has been two billion years since they were free-living unicellular eukaryotes. Their DNA still contains ancient survival programs that protected their single-celled ancestors during stressful times eons ago. When the cell senses a stress, it takes preemptive steps to deal with even greater levels of that stress. In effect, it switches from a "Happy days!" mode to a "Hard times" gear. In future chapters, we'll learn how we can activate this ancient program to our great benefit.

References and Further Reading

K Dokladny et al (2015). Heat shock response and autophagy: cooperation and control. *Autophagy*, 11(2): 200–13.

CL Klaips et al (2018). Pathways of cellular proteostasis in aging and disease. *Journal of Cell Biology*, 217: 51–63.

JJ Reid et al (2019). Energetic Stress and Proteodynamics in Aging and Longevity. *The Science of Hormesis in Health and Longevity.* Chapter 16. Eds S Rattan and M Kyriazi. Academic Press.

11

Autophagy

"Life is an equilibrium state between
synthesis and degradation of proteins."

YOSHINORI OHSUMI

JUST IMAGINE. You're a lonely bacterium swimming around in the ocean billions of years ago. You've evolved to do two things: grow and divide. To perform those assigned tasks, you have to find building materials and sources of energy—in other words, food! Just like your eventual human successors, you need to take in amino acids (the building blocks of proteins), fatty acids (the components of membranes), and sugars (the constituents of carbohydrates and your main energy source). As you swim around, you occasionally encounter these tasty morsels that are mostly remnants of decaying organisms.

But the ocean is a vast place. What if you don't happen to find any of these organic molecules for a few hours, or even a few days? You can certainly try to slow your pace and conserve energy, but even just living requires lots of energy. Think of life as a battle against disorder, or *entropy*, as we called it in high school chemistry. That's the idea that, left to themselves, things will become messier and messier. Just think about your college dorm room. You had to invest energy to maintain order. Your bacterial self has to do the same. So, even in a period of starvation, bacteria—and cells in general—must continue to generate energy and use it to stave off the forces of entropy. But if you can't find food, where does the energy come from?

Luckily, you have lots of spare parts lying around that you can break down and burn, quite literally. A bacterium is more complex than you can imagine, each one containing millions of proteins and other large molecules. The proteins in particular serve as tiny machines, moving cargo around the cell, building new membranes, and even copying your DNA (your genetic blueprint) when it's time to divide. And there's always tons of housekeeping to do. DNA in particular is always breaking, and your protein machines have to be scanning it constantly and patching it up as needed. But just like tractors on a farm, these protein machines themselves get old, wear out, and need to be repaired and, eventually, replaced.

When times are good, you don't worry about repairing or disposing of old components. You just shove them into a corner and build new ones. Have you ever seen an ancient, rusted tractor sitting in the corner of a farmer's field? The farmer bought a new one and hasn't gotten around to getting rid of the old one yet. You're in the same situation. Until recently, you had been finding tons of food, so you just built new components when you needed to and didn't worry about the old ones. But your luck has run out. You haven't encountered a single bite of food for days, and you're starting to get desperate. Something "clicks," and you go into a different gear—survival mode.

You become extra-focused on scanning the ocean for the food you so badly need. But you also start recycling. You send protein machines to scurry around inside you to look for old, unneeded components and slap a "To be recycled" tag on them. You then send other proteins around to look for anything with this tag and drag it off to an internal incinerator, where it will be broken down for scrap and the leftovers burnt to generate some energy. You'll use some of this energy to build new components and some just to stay warm in the cold ocean.

As you can see, the net effect of all this is that even though you're having a hard time finding food, you're actually cleaning up your act on the inside. You're getting leaner and more efficient. Your internal machines no longer have to work around old, discarded components and can do their jobs more easily. And by getting rid of the old components, on average your protein machines and other parts become younger, stronger, and more effective.

When you eventually find food again, you'll go back to your old profligate ways—into kind of a "Happy days are here again!" mode. You'll stop recycling and just let the old stuff sit around again. You'll get a little sloppier when you build new components, because you can always just discard them and start over if you need to. Your main goal now is to begin to make two of everything so you can grow and divide.

A Nobel Prize–Winning Discovery

Let's fast-forward a few billion years now. Cells have learned to live together in large colonies that we call plants and animals. Different groups of cells specialize for different functions—breathing in oxygen, breaking down food, etc. But the two ancient modes—survival and happy days—are still encoded in the DNA that was passed down from every cell's bacterial forebears. These primordial programs still cause a cell to either hunker down and recycle during lean times or grow and divide during periods of plenty. In fact, Japanese scientist Yoshinori Ohsumi won the Nobel Prize in 2016 for discovering this survival mode, which carries the formal name of *autophagy*, or "self-eating."

The discovery of autophagy helped explain some unexpected findings that scientists had made through the years. In 1935, Clive McCay discovered that restricting the number of calories an animal could consume without pushing it into malnutrition caused it to live longer. Others noticed that calorically restricted animals also developed fewer cancers and other diseases as they aged. They wondered what the heck could be going on. How could limiting the amount of nutrition available to a creature make it healthier and increase its lifespan? These strange results were largely ignored until the 1980s and 1990s, when the same thing was observed in monkeys and other primates. This culminated with a report in 2009 by a group at the University of Wisconsin that showed 80 percent of calorically restricted monkeys were still alive at a time when half of normally fed monkeys had died.

Along the way, David Sinclair at Harvard and others showed the same effect in very primitive organisms such as yeast and worms. Clearly something extremely ancient was at play. Yeast in particular were valuable experimental models because, like bacteria, they live as individual cells. Over the course of the last twenty years, scientists have slowly unraveled what's going on right down to the level of genes and proteins.

What Doesn't Kill You Makes You Stronger

Before we dive in, let's remember that genes are DNA recipes for how to make a protein. Human DNA has about twenty thousand such protein-recipe genes. Examples include the gene for hemo-globin, a protein that carries oxygen from our lungs to our tissues, and the BRCA1/BRCA2 genes that contain the recipes for protein machines that repair DNA breaks. It turns out that there are a handful of genes that control the ancient survival program that we talked about earlier—the one that jumpstarts recycling and the process of hunkering down to wait for better times.

In the chapter on cell signaling, we talked about the two main nutritional sensors in the cell, AMPK and mTOR. Recall that AMPK is kind of a low-energy alert that is active when the cell starts to run low on ATP, such as during fasting periods. mTOR is the gateway to protein synthesis and overall cell growth and is activated in times of plenty—lots of ATP and amino acids.

So far so good. We see how the cell can sense the levels of fuel and building blocks available to it. What happens from there? The next important players are called SIRT1 and SIRT3. Both of these proteins are responsible for activating survival programs in different parts of the cell. SIRT1 sends an alert to the nucleus that it's time to go into survival mode. This turns on the production of protective genes that make antioxidants and other molecules to help the cell survive during times of nutritional deprivation. SIRT3 activates a similar program in the cell's mitochondria.

Recall that mitochondria are actually the remnants of a long-ago event when one type of bacteria started living inside another type. This might have begun as a parasitic relationship, but over time it worked to the mutual advantage of both the host cell and its guests. Mitochondria even still have their own DNA, betraying their bacterial past. SIRT3 sends the survival signal to these internal "organisms," letting them know that times are tough. When food is scarce, cells rely even more heavily on their mitochondrial house guests to produce energy, especially from fat.

More on that subject later. But in anticipation of forcing the mitochondria to run hot, SIRT3 gives them a heads-up that they're going to need to work harder. This allows mitochondria to start making antioxidants and other molecules to help them deal with the side effects of ramping up energy production.

The way that SIRT1 and SIRT3 (examples of a class of proteins called sirtuins) operate is pretty ingenious. Whether a cell is breaking down sugar or fat to produce energy, it ends up with a two-carbon molecule called an acetyl group loaded into a tiny truck called coenzyme A. You may vaguely remember the term acetyl-CoA from your distant past. When acetyl-CoA is plentiful, it means that lots of fuel molecules are being broken down. These abundant acetyl groups get attached to proteins to turn them on or off, thus activating the "Happy Days!" program. However, SIRT1 and SIRT3 reverse this program by removing acetyl groups from proteins (and from DNA as well). This deactivates the "Happy Days!" mode and switches the cell into survival gear.

We can think of these acetyl groups as pieces of fuel molecules, so the food itself is attached to proteins and DNA to determine whether the cell will party hardy or hunker down. David Sinclair, mentioned earlier, discovered that SIRT1 and SIRT3 are also activated by plant molecules such as the resveratrol found in grapes and red wine. The bright colors in many berries, fruits, and vegetables fall into the same category called polyphenols, which is why these substances are so good for us. They actually mimic some of the effects of caloric restriction, even when we're eating plenty, by activating the sirtuins.

As we've previously explored, this notion of sensing impending hard times and taking steps to get ready is referred to as hormesis. The phrase "What doesn't kill you makes you stronger" gives you a good sense of what's going on. Many stressors activate this same survival mode, including:

- Exercise
- Polyphenols and other plant molecules
- Cold
- Heat
- High altitude
- Fasting and caloric restriction

Even low levels of radiation provide the same sort of stress. Believe it or not, there are actually people who go to the bottom of abandoned mine shafts to intentionally expose themselves to low levels of radiation to trigger this effect. Saunas are popular in Scandinavia and have been proved to improve health by activating hormesis. The pigments in brightly colored berries mentioned above are actually poisons produced by the berries to protect against insects. When we eat these berries, they have a mild toxic effect that once again serves as a small stressor that makes our cells think there's trouble and causes them to activate their survival program.

How does all this relate to autophagy? Well, that recycling process is just one part of the ancient survival program our cells inherited from bacteria. In times of plenty, when mTOR is activated, autophagy is turned off. Why recycle when there's lots of everything you need? Just like many of us, cells don't recycle unless they're forced to. However, once mTOR has been turned off for a while, cells start to worry. How long before food will be available again? Hours, days, weeks? Our cells have no idea, so they have to plan for the worst. And a big part of that survival planning involves autophagy.

The main controller of autophagy is a family of proteins called the FOXOs—we'll just use FOXO for simplicity. In times of nutritional abundance, mTOR is active and slaps a tag on FOXO that prevents it from going into the nucleus. When mTOR is turned off, this tag doesn't get applied, and FOXO marches right into the DNA storage vault and turns on a set of genes that activate part of the survival program, including autophagy. And just to emphasize how everything is interconnected, it turns out that AMPK and SIRT1 help activate FOXO as well.

Fig. 87: Caloric restriction

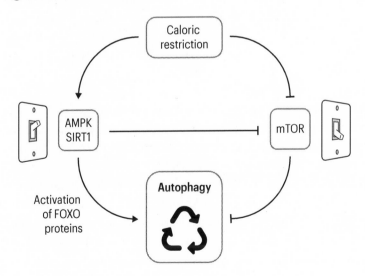

How Does Autophagy Work?

So how does the autophagy process itself work? There are actually three different forms, which we can think of as:

1 One protein at a time
2 A small "bite" of part of the cell
3 A big "bite" of part of the cell

The most granular of these is referred to as *chaperone-mediated autophagy*. Recall that a chaperone is an important kind of protein that either helps other proteins fold into the right shape or escorts them to the incinerator if they can't fold properly or are no longer needed. Chaperone-mediated autophagy is extremely important for the correct operation of the cell. Without it, some proteins continue to send signals when they shouldn't and otherwise cause the cell to operate inefficiently. Only birds and mammals have this type of autophagy, so it appears to be a more recent invention of evolution.

"Small bite" autophagy has the formal name of *micro-autophagy*. Imagine walking around your bedroom with a small bag and just grabbing anything that will fit, putting it inside, and taking the bag to the recycling bin. This is what happens in micro-autophagy. The cell just takes small bites of its own internal contents and recycles them. This provides a sort of gradual renewal. Old proteins and other components are gradually broken down, which prompts the construction of new ones to replace them.

"Big bite" autophagy is known as *macro-autophagy*. As the name implies, this is like walking around your place with a big garbage bag, stuffing anything inside it that will fit, and taking it to a recycling center. For example, you might grab the microwave in the corner and get rid of it. Now this might sound pretty wasteful, but the process is a bit more discriminating than this paragraph implies. Let's imagine that your roommate occasionally puts Post-it Notes on old stuff that needs to be recycled. When you walk around the apartment with your garbage bag, you put in it only the things that have Post-its stuck to them.

Fig. 88: Macro- and micro-autophagy

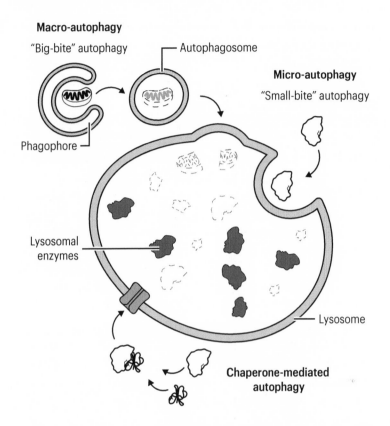

Macro-autophagy

"Big-bite" autophagy

Autophagosome

Micro-autophagy

"Small-bite" autophagy

Phagophore

Lysosomal enzymes

Lysosome

Chaperone-mediated autophagy

One extremely important sub-category of macro-autophagy is also called *mitophagy* because it focuses on recycling mitochondria. For a physical analogy, imagine you have space heaters around your apartment. As those heaters get old, they become less efficient, producing less heat and consuming more electricity to do so. They might even start to give off bad odors as they get leaky. The lazy approach would be to just go out and buy new space heaters and let the old ones continue to operate. This might heat up your apartment, but it would drive up your electric bill, clutter the apartment, and make the whole place start to smell pretty bad. You can see how it would be far better for your roommate to periodically inspect the heaters and label the ones that should be replaced. You would then go around with your garbage bag once in a while, load in the old space heaters, and take them to the recycling center. Of course, you'd still have to buy new ones to replace them, but at least your apartment would operate efficiently—and with no clutter or noxious odors!

All three types of autophagy work in much the same way. An isolation membrane called a vesicle is put around an area of the cell. When complete, this vesicle is called an *autophagosome*. It then fuses with an acid-filled compartment called a *lysosome*. The contents are broken down and made available for reuse.

Fig. 89: Lysosomes

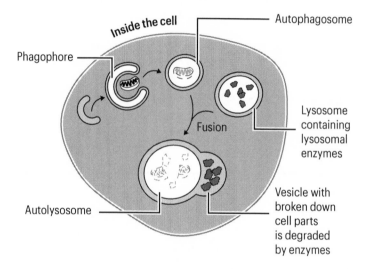

With advancing age, lysosomes become less functional and less able to break down the proteins and other contents captured in the autophagosome. Instead, proteins and other components begin to form clumps in lysosomes, resulting in deposits called lipofuscin. When they occur in the skin, these are referred to commonly as "liver spots" or "age spots."

Free Radicals or Reactive Oxygen Species (ROS)

The analogy of mitochondria as space heaters is actually pretty apt. Space heaters and mitochondria have the following in common:

* Both have the main job of producing energy—heat in the case of a space heater and ATP in the case of mitochondria.

* Both have to consume one energy source in order to produce a different kind of energy. Space heaters plug into the wall and use electricity to generate heat. Mitochondria take in fuel molecules broken down from glucose and fat and use them to generate ATP.

* Both can get leaky as they age. This can cause them to become less efficient and to generate noxious by-products. In the case of space heaters, this can be foul fumes. Old mitochondria leak nasty free radicals.

* Both need to be recycled and replaced when they get old.

We've already mentioned free radicals. They sound like a bunch of dangerous anarchists, and they are. As their other name— reactive oxygen species, or ROS—suggests, they mainly involve different forms of oxygen. Free radicals principally exist in three forms—superoxide, hydrogen peroxide, and hydroxyl radical. Superoxide and hydroxyl radical are the most reactive and thus have the most short-term danger. However, hydrogen peroxide is sneaky. Because it is more stable, it doesn't pose as much of an immediate threat. But that very stability means that hydrogen peroxide can travel around in the cell, potentially reaching the nucleus and damaging the precious DNA stored there. This is why it's so important to break down old mitochondria and replace them with more efficient new ones.

Cells have that special version of "big bite" autophagy specifically designed to recycle mitochondria, mentioned earlier— mitophagy. Recall that mitochondria were once free-living bacteria. Like all bacteria, they maintain a charge difference across their inner membrane—more negative inside and more positive outside. As we learned earlier, they use this charge difference to generate ATP. As mitochondria get old and damaged, this charge difference decreases. The cell monitors the charge across mitochondrial membranes and tags mitochondria for destruction when it drops below a certain threshold.

We mentioned two of the primary proteins involved in this process earlier when discussing Parkinson's disease: PINK1 and Parkin. In a mitochondrion with a normal membrane charge difference (also called potential), PINK1 gets imported through a pore into the interior, where it is degraded. When the membrane potential falls as the mitochondrion gets old, PINK1 builds up on the surface of the mitochondria. This attracts Parkin, which labels the mitochondrion for destruction in a nearby lysosome. Mutated versions of the genes for PINK1 and Parkin are associated with versions of Parkinson's that run in families. A malformed version of PINK1 may fail to recognize an old mitochondrion that needs to be recycled. A bad version of Parkin may not tag such a mitochondrion even though it has PINK1 all over it.

Medical Implications

Autophagy plays an important role in various diseases. Researchers are exploring the benefits of both increasing autophagy as well as interfering with it. Cancer illustrates the double-sided nature of the process. Autophagy helps prevent cancer by getting rid of the misfolded proteins and leaky mitochondria that can contribute to DNA damage and the initiation of cancer. However, once cancer has developed, many cancer cells can use autophagy to help deal with the increased stress that comes from rapid proliferation. But even in the case of established cancer, inhibiting autophagy can sometimes be beneficial. Without autophagy, cancer cells often lose important stress protection and may be more vulnerable to chemotherapy and other treatments.

Scientists are evaluating the possibility of developing drugs capable of either stimulating autophagy or inhibiting it. As mentioned above, inhibiting autophagy may rob cancer cells of an important defensive measure; however, doing so would also adversely affect normal cells. Medications to stimulate autophagy might be very useful in diseases involving protein aggregation, such as Alzheimer's and Parkinson's. However, any drug must compete with the very simple behavioral changes that are already known to safely increase the rate of autophagy, including exercise and intermittent fasting. Older drugs such as valproic acid could also be repurposed as autophagy inducers. One drug under active investigation is rapamycin, a potent inhibitor of mTOR. mTOR effectively shuts down autophagy, so inhibiting mTOR is a way of increasing the rate of this recycling process.

Activating autophagy may not always be beneficial. In some people, lysosomes are less active than normal, reducing their ability to dissolve the cargo delivered to them in autophagosomes during autophagy. This is true in many cases of Parkinson's and Alzheimer's. Just like with a blocked pipe, pushing more down the drain may just make problems worse, so simply increasing the rate of autophagy may not be enough.

The "dissolving power" of lysosomes also decreases with age. Scientists are now investigating compounds that can increase lysosomal activity. Lysosome production and function are controlled by a master transcription factor called TFEB. Fortunately, there are natural substances that also increase the activity of TFEB and hence lysosomes. Chief among these is the phytochemical curcumin found in turmeric. Another is the molecule resveratrol found in grapes and red wine. Aspirin, too, is able to increase the transcription of the gene for TFEB, resulting in an increase in the transcription factor's concentration. Our old friend SIRT1 also plays a role by removing acetyl groups from TFEB, thereby increasing its ability to turn on the genes involved in constructing new lysosomes.

A recent study also suggests that while activating autophagy early in the course of neurodegenerative diseases like Alzheimer's and Parkinson's may be beneficial, doing so late may do more harm than good. It appears that mTOR is initially overactive in these diseases, thereby turning off autophagy. However, as the misfolded proteins accumulate and the cell becomes oxidized, there's a tipping point at which other pathways override mTOR and turn on autophagy. Because the neuron is so damaged at this point, the recycling triggered by autophagy ends up killing the cell. This emphasizes the fact that the sooner steps are taken to reduce mTOR and increase autophagy, the better.

Summary

As we've seen in earlier chapters, protein misfolding increases with age and can lead to a number of deadly diseases including Alzheimer's and Parkinson's. Biotech companies are now working on small molecules that might increase the overall rate of autophagy. The hope is that the increased activity will activate autophagy and thereby clean up some of this protein mess before it gets toxic enough to permanently damage or even kill cells undergoing stress from misfolded proteins. Fortunately for us, we don't have to wait for the development of such medications. This book is packed full of ways that you can generate a small stress capable of ramping up autophagy. And healthy plants contain substances like sulforaphane that can also increase the rate of autophagy. The easiest and most powerful way to force your cells to start recycling is by fasting for sixteen hours or longer. As we'll see in the upcoming chapter on intermittent fasting, going without calories for a few hours acts as a potent signal to the cell to activate its ancient autophagy program.

References and Further Reading

MC Barbosa, RA Grosso, and CM Fader (2019). Hallmarks of aging: an autophagic perspective. *Frontiers in Endocrinology*, 9 January 2019.

I Dikic and Z Elazar (2018). Mechanism and medical implications of mammalian autophagy. *Nature Reviews Molecular Cell Biology*, 19: 349–64.

KA Escobar et al (2018). Autophagy and aging: maintaining the proteome through exercise and caloric restriction. *Aging Cell*, 18(1): e12876.

J Ren and Y Zhang (2015). Targeting autophagy in aging and aging-related cardiovascular diseases. *Trends in Pharmacological Sciences*, 39(12): 1064–76.

12

Plants and Phytochemicals

"One of the biggest scientific discoveries made is that colorful fruits and vegetables contain many disease-fighting compounds known as phytochemicals and that we need the protective benefits of the full spectrum of their bright colors."

JAMES JOSEPH

IN THIS chapter, we'll try to understand why eating a variety of plants is such an important step for optimizing our health. "Phyto" comes from the Greek word for plant. Phytochemicals are chemicals produced by plants for defense against bacteria, viruses, fungi, insects, and even animals. They are often poisons for the organisms consuming them. In a few cases, these toxins are toxic enough to kill an unwary organism. Even milder versions often have strong effects on the consumer, leading such plants to be the mainstays of traditional medicine for thousands of years. Salicin, for instance, is a potent anti-inflammatory found in willow bark that led to the development of aspirin. The English yew tree is the source of an anti-cancer agent called paclitaxel.

Western medicine is just catching on to the therapeutic value of phytochemicals.

Plants produce a wide range of chemical compounds for many purposes. Some chemicals make leaves and berries brightly colored to attract or repel different organisms. Such chemicals act as dyes and often have powerful physiological effects when eaten. Other phytochemicals act as toxins to ward off insects and pathogens. Phytochemicals are often produced by plants in response to various stressors (heat, cold, drought, etc.), which is why they are referred to as stress chemicals. Phytochemicals can have a number of effects on humans and other mammals, including:

- Stimulating antioxidant activity
- Inhibiting the growth of harmful microbes
- Increasing production of detoxification enzymes
- Decreasing platelet aggregation
- Modulating the immune system
- Promoting anti-cancer activity

Because of these effects, many phytochemicals can play a role in preventing or treating various diseases. Medicinal plants have been used since antiquity but were largely discontinued in the Western world during the twentieth century. Biochemists were able to isolate specific active compounds from plants and produce them in pure, synthetic forms. These had the advantage of strength, specificity, and reproducibility. However, the direct use of plants has been rediscovered in the last few decades with the realization that some medicinal plants are just as effective as synthetic medicines, often with fewer side effects and at a far lower cost.

Phytochemicals fall into four main categories—phenols, terpenoids, thiols, and alkaloids. We'll look closely at their structures as well as their functions. Feel free to ignore the chemical drawings and formulae if you like. They are included for completeness but are not necessary to appreciate the importance of healthy plants and their products.

Phenols

Phenols are the largest group of phytochemicals, with members spread throughout the plant kingdom. A phenol is a six-carbon ring with a hydroxyl group (-OH) attached. Figure 90 shows three different ways of representing a phenol. In the first and third illustrations, hydrogen atoms are ignored and carbon atoms are represented by corners where two lines come together.

Fig. 90: Phenols

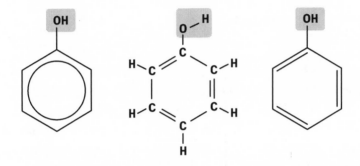

The six-carbon ring (also known as a benzene ring) is very hydrophobic, while the hydroxyl group is hydrophilic. So phenols can be a bit schizophrenic, with parts that like water and parts that don't. One special aspect of the benzene ring is that it allows the electrons in the various carbon-carbon bonds to be spread across the entire ring as shown in figure 91.

This ability to share the electrons across the entire ring means that such a ring can easily handle an extra electron or get by with one fewer. Such molecules make good antioxidants because they can either take the unpaired electron from a free radical like superoxide (O_2^{*-}) or donate one to it to form hydrogen peroxide (H_2O_2). Most commonly, these molecules donate the hydrogen atom (proton plus electron) from the hydroxyl group (-OH) attached to the phenol ring in order to quench a free radical on another molecule.

Fig. 91: Six-carbon ring

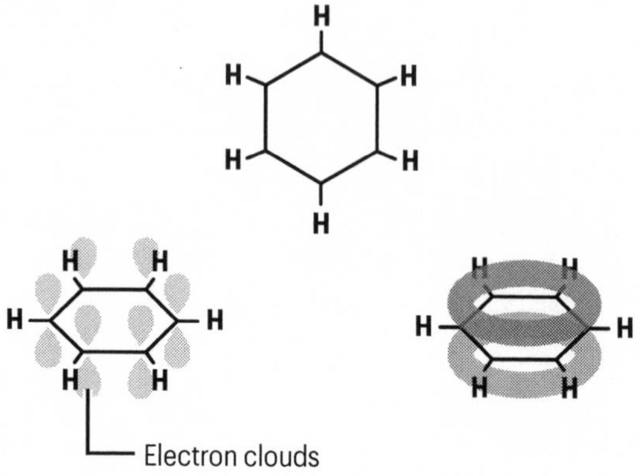

Electron clouds

As their name implies, polyphenols often have multiple phenol groups and thus multiple benzene rings, making them even more powerful antioxidants. For example, you might have heard of the "red wine molecule" called resveratrol. It is structured with two phenol groups (one with a second hydroxyl group). In figure 92, we see how resveratrol can donate a hydrogen atom (proton plus electron) to a dangerous DPPH molecule to neutralize its unpaired electron.

Fig. 92: Phenol groups

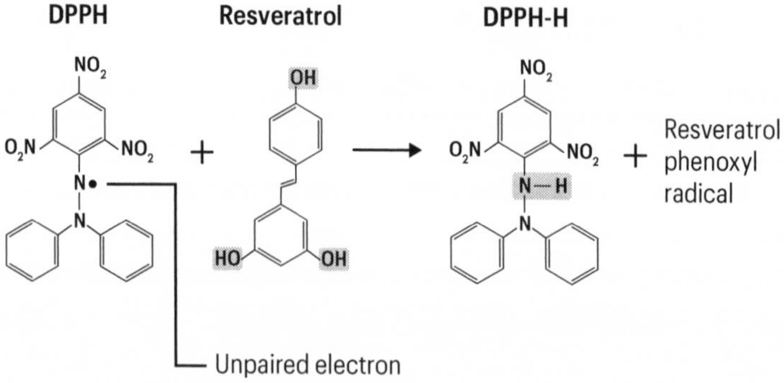

DPPH   Resveratrol   DPPH-H

Resveratrol phenoxyl radical

Unpaired electron

Now the resveratrol molecule takes the unpaired electron. But that electron is able to spread over the entire benzene ring of resveratrol and is much less reactive.

As figure 93 illustrates, phenols fall into several subcategories, including:

* Phenolic acids
* Stilbenes (such as resveratrol)
* Flavonoids
* Coumarins
* Lignans

Fig. 93: Phenol chart, foods

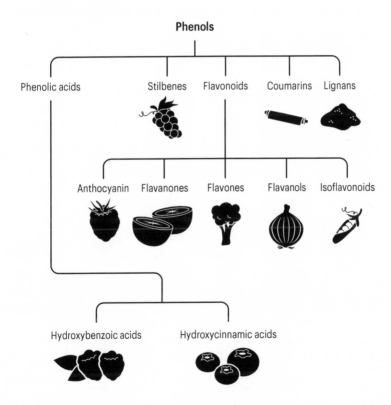

All of these compounds are abundant in fruits, vegetables, herbs, and spices. They are known to have both antioxidant and anti-inflammatory properties.

Phenolic Acids

A phenolic acid is a phenol with a carboxylic acid functional group. The classic examples—hydroxybenzoic acid and hydroxycinnamic acid—are shown in figure 94. Hydroxybenzoic acid has the -COOH group directly attached to the ring, while hydroxycinnamic acid has a two-carbon spacer. They are valued for their antioxidant properties. You'll often see hydroxybenzoic acid listed as a preservative in foods because of its ability to fight oxidation. It is found in olives and most fruits, especially berries. Hydroxycinnamic acids are found in tea leaves, coffee, red wine, red fruit, vegetables, and whole grains.

Fig. 94: Hydroxybenzoic and hydroxycinnamic acid

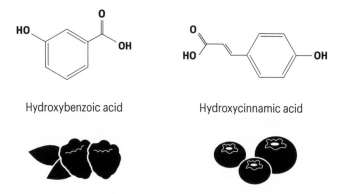

Hydroxybenzoic acid   Hydroxycinnamic acid

In addition to their antioxidant properties, these molecules are able to bind metals like copper or iron that can be dangerous when freely roaming the cell.

Stilbenes

Stilbenes are molecules with two phenol groups connected by a two-carbon bridge with the carbons double-bonded together (ethylene). Resveratrol is the poster child for this type of phenol. A wide variety of plants produce stilbenes and use them to fight off fungi, nematodes, and herbivores. Plants also produce stilbenes in response to various stressors such as UV light, drought, heat, and exposure to heavy metals, so we can think of them as stress signals generated by plants. Not surprisingly, these molecules may induce small stressors in our own cells, leading to the protective responses we see in hormesis.

Fig. 95 : Resveratrol

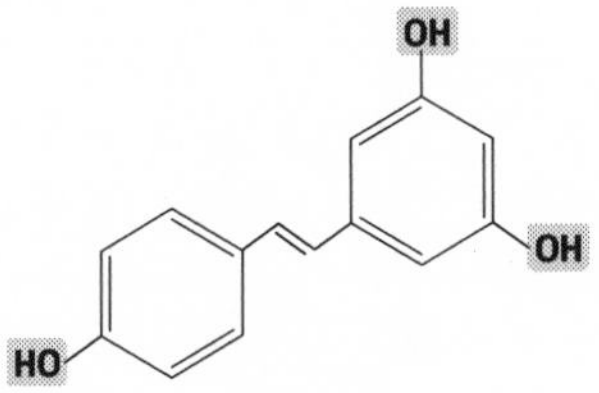

Resveratrol

Stilbenes are found in grapes, berries, and certain tree barks. They are being investigated for use in several diseases of aging, including neurodegenerative disorders such as Alzheimer's and Parkinson's. Stilbenes are able to activate the NRF2 stress defense pathway via cAMP and AMPK signaling. This provides protection against oxidative stress. Stilbenes have been found to have therapeutic value in the following conditions:

Obesity. Resveratrol can activate AMPK, thereby helping improve insulin sensitivity. Stilbenes can also help counteract the inflammation that accompanies obesity. These compounds also seem to help increase fat metabolism.

Type 2 diabetes. Resveratrol and other stilbenes increase insertion of GLUT4 transporters into membranes. They also reduce some of the inflammation associated with high levels of glucose.

Macular degeneration and neurodegenerative disorders. High levels of oxidative stress are involved in macular degeneration (in the eye) and neurodegenerative diseases such as Alzheimer's and

Parkinson's. Stilbenes can greatly reduce the oxidative load as well as chelate (bind) the free iron that can generate the dangerous hydroxyl radical. Another benefit of stilbenes is the increased rate of autophagy in these diseases that all involve the accumulation of misfolded proteins.

Fig. 96 : Stilbenes

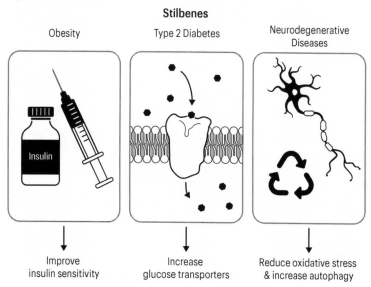

Stilbenes

Obesity	Type 2 Diabetes	Neurodegenerative Diseases
Improve insulin sensitivity	Increase glucose transporters	Reduce oxidative stress & increase autophagy

Flavonoids

The flavonoids are perhaps the largest and most diverse category of phenols. More than four thousand different types of flavonoids have been found in fruits, vegetables, and drinks such as tea and coffee. The three main subcategories of flavonoids look like this:

Fig. 97: Flavanoids

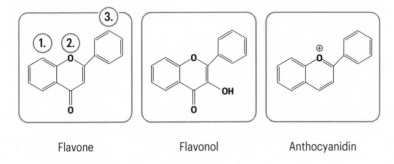

| Flavone | Flavonol | Anthocyanidin |

The main structural features are:

1   A six-carbon benzene ring
2   Another six-element ring featuring an oxygen atom
3   Another benzene ring connected via a carbon-carbon bond to the oxygen-containing ring

There are many variations of flavonoids. Flavones are ketones with an oxygen double-bonded to the middle ring. Flavonols also have a hydroxyl group (-OH) attached to the middle ring. In anthocyanidins, the oxygen in the middle ring carries a positive charge because it is doubly bonded to a carbon atom in that ring.

Flavonoids have a broad range of activities and are able to exert antimicrobial, cytotoxic, anti-inflammatory, and anti-tumor effects. But their most studied role is as antioxidants, protecting cellular components such as carbohydrates, proteins, lipids, and DNA against assault by oxygen. The following table lists some of the most common flavonoids, their effects, and which plants tend to contain them.

Flavonoid	Category	Actions	Sources
Luteolin	Flavone	Decreases blood blood pressure, anti-inflammatory, anti-cancer	Leaves, rinds, barks, clover, ragweed, celery, parsley, broccoli, onion, carrots, apple skins
Apigenin	Flavone	Antioxidant, anti-inflammatory, anti-cancer (cell cycle arrest, apoptosis), neuroprotective, sleep-promoting	Parsley, celery, onions, oranges, herbs (chamomile, thyme, oregano, basil), drinks (tea, beer, wine)
Tangeritin	Flavone	Antiproliferative, anti-invasive, antimetastatic, antioxidant, stops or delays cell cycle	Tangerine and other citrus peels
Fisetin	Flavonol	Antioxidant, senolytic, anti-inflammatory	Strawberries, apples, onions
Cyanidin	Antho-cyanidin	Antioxidant	Blackberries, elderberries, red cabbage, raspberries
Malvidin	Antho-cyanidin	Antioxidant	Blueberries, grapes
Epigallo-catechin gallate (EGCG)	Flavonol	Antioxidant, anti-cancer	Tea (especially green)
Epicatechin	Flavonol	Antioxidant, anti-cancer	Chocolate, apples
Genistein	Isoflavone	Antioxidant, anti-cancer	Soy

Proanthocyanidins are polymers (chains) of flavonoids. Rich sources include chocolate, cinnamon, nuts, and cranberries.

Lignans

Lignans are another type of polyphenol, found in fiber-rich plants such as flax. Plant lignans are converted to one of the following compounds in the gut: enterodiol, enterolactone, or estradiol. Lignans have a strongly anti-tumor effect, especially for breast, colon, and prostate cancer.

Fig. 98: Lignans

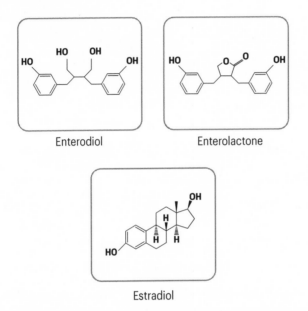

Enterodiol

Enterolactone

Estradiol

In addition to being polyphenols with all the benefits that come from that chemical type, lignans are also phytoestrogens—plant compounds capable of acting as weak ligands for estrogen receptors. This may explain some of the benefit in cancers that depend on steroid signaling (breast and prostate). Lignans may be similar enough to estrogen to bind weakly to estrogen receptors and prevent the estrogen molecule from doing so. This decreases the amount of proliferative signaling being delivered to these cells.

Terpenoids

Terpenoids are plant compounds made from isoprene building blocks. Isoprene is a branched five-carbon molecule with two carbon-carbon double bonds, as shown in figure 99a. Two isoprenes can be combined to form myrcene, which can be used to form the other terpenoids, menthol and limonene, as shown.

Fig. 99a: Terpenoids

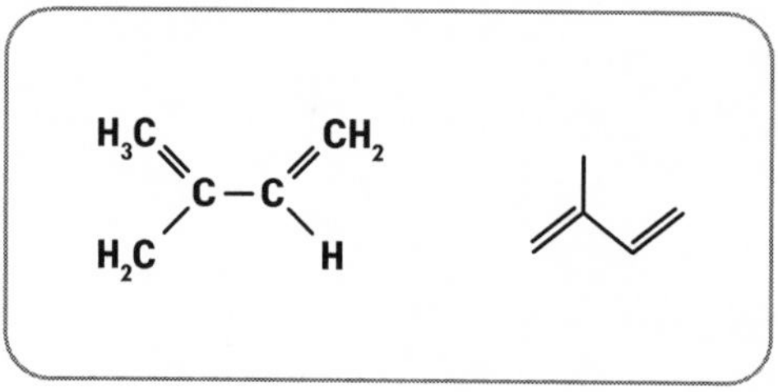

Isoprene

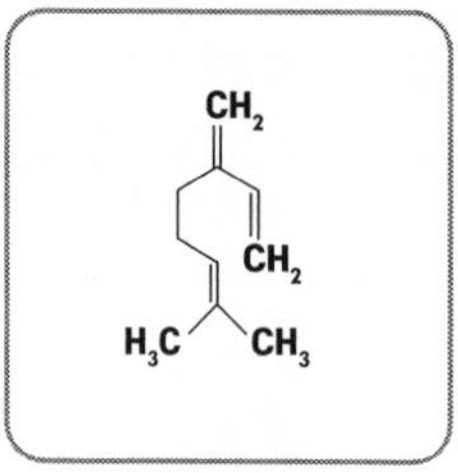

Myrcene 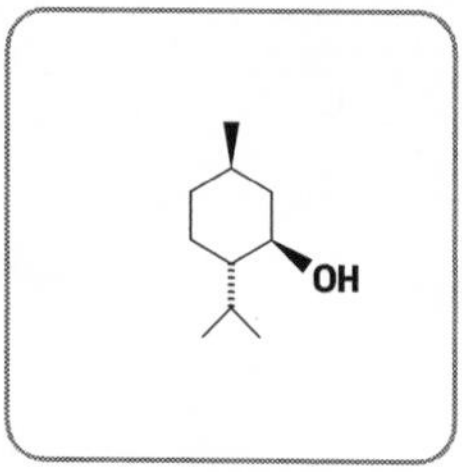   Menthol 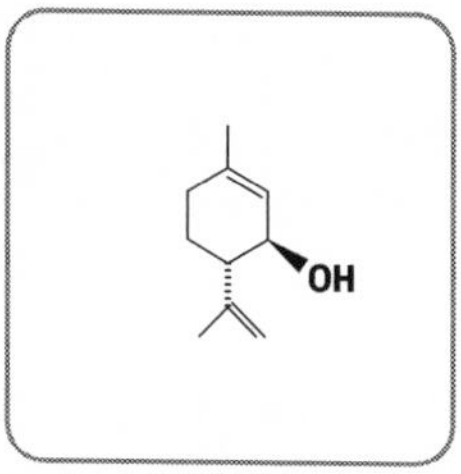   Limonene

Terpenoids can contain multiple five- or six-member rings connected by chains of isoprene units, as seen in lanosterol.

Fig. 99b: Terpenoids

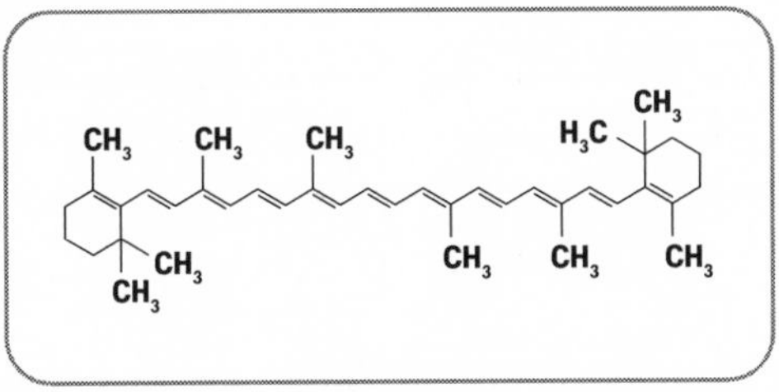

ß-Carotene

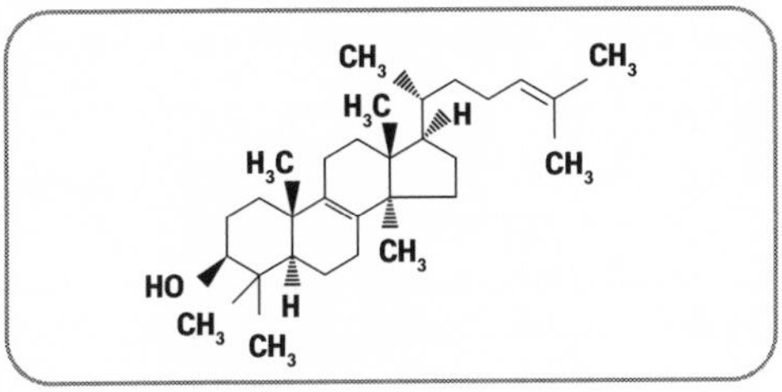

Lanosterol

The terpenoids are natural lipids found in just about every class of living thing. These molecules can contain oxygen atoms and others but are mostly hydrocarbons and mostly lipophilic. This allows them to pass relatively easily through cell membranes. Many of them have strong aromas or flavors (e.g., menthol). Terpenoids are often found as essential oils in plants. The following table provides some examples.

Name	Actions	Sources
Eucalyptol	Anti-inflammatory via cytokine inhibition, decreases mucus, vasodilation	Eucalyptus oil, wormwood, rosemary
Limonene	Anti-inflammatory, antioxidant, anti-stress, reduces triglycerides	Trees, citrus peels

Name	Actions	Sources
Camphor	Anti-inflammatory, pain relief, decongestant	Camphor trees
Pinene	Anti-inflammatory, pain relief, anxiolytic, bronchodilator	Pine needles, rosemary, basil, cannabis
Retinol	Antioxidant, night vision	Conversion from ß- carotene
Lanosterol	Precursor for steroids	Meat
Squalene	Precursor for steroids	Wheat germ, olives
ß-Carotene	Precursor for retinol, antioxidant	Carrots, onions, peas, spinach, squash, pumpkin
Lycopene	Antioxidant, anti-cancer	Tomatoes, watermelon, grapefruit
Canna-binoids	Pain relief, anxiolytic	Cannabis
Ginkgolide	Migraine relief, reduce platelet aggregation	Ginkgo biloba
Curcu-minoids	Neuroprotective, antioxidant, anti-tumor, cardio-protective	Turmeric, mustard seed

Curcuminoids

Many of the various types of phytochemicals found in turmeric fall into the category of curcuminoids (e.g., cumin). As shown in figure 100, these molecules have two phenol rings, each with a hydroxymethyl group. The rings are connected by a seven-carbon bridge with two ketone groups and two carbon-carbon double bonds. Curcumin can switch back and forth between a keto form and an enol form in which it has accepted a hydrogen atom to reduce one of the ketone oxygens.

Curcumin has a number of important biological activities that affect just about every organ system in the body. In addition to its ability to protect neurons against oxidative stress, the molecule also has strong anti-inflammatory properties and is used to help with the symptoms of arthritis. Curcumin also helps activate autophagy.

Fig. 100: Keto & enol forms of curcuminoids

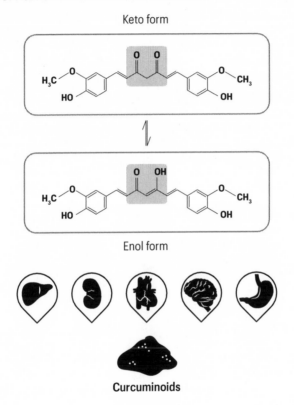

Keto form

Enol form

Curcuminoids

Thiols

Thiols are sulfur-containing compounds that often have strong odors similar to garlic or rotten eggs. Plants often produce them in response to different types of stress—drought, temperature extremes, etc. One of the most important of these is *glutathione*. Glutathione is a tripeptide consisting of glutamate, cysteine, and glycine. It is also the most abundant thiol found in animal cells and serves as a critical antioxidant. Glutathione exists in two different forms. When reduced (electron-rich), the cysteine has a hydrogen atom on its sulfur, as shown below. This form is often abbreviated as GSH or G—SH. In oxidizing conditions, two glutathione molecules link together by means of a disulfide bridge between the cysteines.

Fig. 101: Thiols

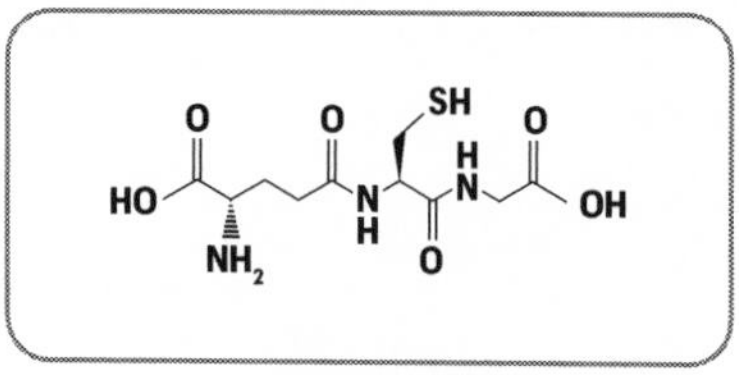

Glutathione

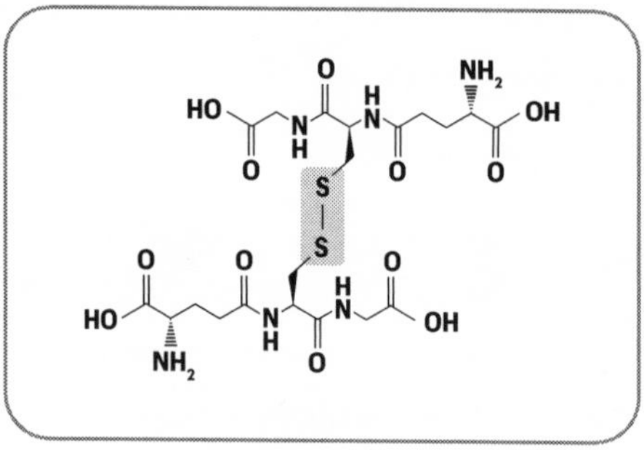

Glutathione in oxidizing conditions

This allows glutathione to act as a sensor that can indicate how much oxidative stress a cell is currently under. Glutathione is an important antioxidant that works with the enzyme glutathione reductase and the metal selenium to turn the reactive oxygen species hydrogen peroxide into water. Depletion of GSH in the hippocampus may be an early sign of Alzheimer's.

Fig. 102: Glutathione chart

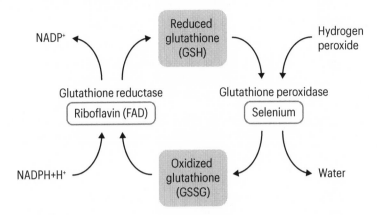

Thiols are also important metal chelators. That is, they can bind to metals such as iron and cadmium to remove them from the body. In effect, they act as sponges to mop up excess metals that can contribute to oxidative stress.

Alkaloids

Alkaloids are bitter-tasting compounds that contain rings incorporating a nitrogen atom. Most of them are bases that form salts with acids. Alkaloid compounds can be found in a wide variety of flowering plants, fungi, and even some animals (e.g., frogs). Alkaloids generally contain one or more of the following ring structures:

Name	Structure	Description
Pyrrolidine		Pentagonal ring with one nitrogen and no double bonds.

Name	Structure	Description
Pyridine		Hexagonal ring with one nitrogen and three double bonds. Analogous to benzene.
Piperidine		Hexagonal ring with one nitrogen and no double bonds. Found in black pepper.
Quinoline		Two hexagonal rings, one benzene and one pyridine. Includes anti-malarials like quinine, chloroquine, and primaquine.
Isoquinoline		Two hexagonal rings very similar to quinoline. The only difference is the position of the nitrogen atom in the heterocyclic ring. Several opioids fall into this category including morphine, codeine, and heroin.
Xanthine		Two rings each with two nitrogen atoms—one hexagonal and the other pentagonal. Includes caffeine (coffee), theophylline (tea), and theobromine (cocoa). Mild stimulants.

Many alkaloids combine multiple ring structures. For example, nicotine contains both a pyrrolidine ring as well as a pyridine ring. Alkaloids have been used for thousands of years as dyes, spices, medications, and poisons. Plants rely on them to help fight off insects and fungi because of their bitter tastes and smells.

Mechanisms
....

Phytochemicals have powerful effects on our health. Many phenols and other phytochemicals are direct antioxidants that can quench free radicals and other reactive oxygen species within cells. But paradoxically, many phytochemicals may act as mild poisons. If you think back to the chapter on hormesis, you'll recall that cells often respond to mild stressors by preparing for even greater stress. In many cases, phytochemicals directly impact mitochondria. So, compounds that plants make to fight off bacterial pests can also affect the mitochondria (former bacteria) within eukaryotic cells. In many cases, this activates a stress response program that causes the cell to ramp up its production of chaperones—proteins that help other proteins fold correctly as well as recycling old proteins. These steps help the cell deal with the current stress but inadvertently reduce the risk of many diseases that stem from the aggregation of old and misfolded proteins, like Alzheimer's, Parkinson's, and many others. The stress response can also lead to the recycling of old mitochondria (mitophagy) and the production of new ones, leaving the cell in a better position to deal with future threats.

The bottom line is that one of the easiest things you can do to improve your health is eat lots of fresh fruits and vegetables. We're finding that what you eat is more important than what you avoid. Instead of counseling patients to give up certain foods that they love, many health professionals are emphasizing the approach of maximizing intake of healthy plants. So don't give up your pizza, fried chicken, and other favorite dishes. But do consume as many fresh fruits and vegetables as possible—ideally at least five servings per day. And given the wide variety of delicious options—from apples to avocados to almonds—it's really not much of a hardship at all.

References and Further Reading

A Amalraj (2016). Biological activities of curcuminoids, other biomolecules from turmeric and their derivatives—a review. *Journal of Traditional and Complementary Medicine*, 7(2): 205–33.

AS Dubrovina and KV Kiseley (2017). Regulation of stilbene biosynthesis in plants. *Planta*, 246: 597–623.

GD Kim et al (2014). Immunosuppressive effects of fisetin against dinitrofluorobenzene-induced atopic dermatitis-like symptoms in NC/Nga mice. *Food and Chemical Toxicology*, 66: 341–9.

J Kurek (2019). *Alkaloids: Their Importance in Nature and for Human Life*. Open access. DOI: 10.5772/intechopen.73336. November 13, 2019.

KB Martinez, JD Mackert, and MK McIntosh (2017). Polyphenols and Intestinal Health. *Nutrition and Functional Foods for Healthy Aging*. Chapter 18. Academic Press.

M Reinisalo et al (2015). Polyphenol stilbenes: molecular mechanisms of defence against oxidative stress and aging-related diseases. *Oxidative Medicine and Cellular Longevity*, 2015; 2015: 340520.

B Salehi et al (2019). The therapeutic potential of apigenin. *International Journal of Molecular Sciences*, 20(6): 1305.

M Saxena et al (2013). Phytochemistry of medicinal plants. *Journal of Pharmacognosy and Phytochemistry*, 1(6): 168–82.

TG Son, S Camandola, and MP Mattson (2008). Hormetic dietary phytochemicals. *Neuromolecular Medicine*, 10: 236–46.

13

Exercise

"True enjoyment comes from activity of the mind
and exercise of the body; the two are ever united."

WILHELM VON HUMBOLDT

O THER THAN stopping smoking, there's nothing better you
can do for your health than exercising. It has often been
said that if exercise were a drug, people would pay a thousand dollars per pill. If you look back over the course of this book,
it will become clear why exercise is so important. As complex
organisms, we tend to switch between two ancient modes that
are programmed into our cells at the deepest level: grow or hunker
down. We enter growth mode when times are good. If the levels of
ATP in our cells are high, and if there are plenty of amino acids to
build new proteins and plenty of glucose to generate more energy,
our cells and bodies are programmed to build their reserves,
repair what's broken, and get ready to reproduce. Remember
that reproduction is always evolution's core imperative. But what
happens if we're past the reproductive phase of our lives? At that
point, growing mainly means building up fat reserves.

The hunker down mode is the one we're really interested in. It's the mode that we seldom switch into in our modern world. In fact, we do everything possible to avoid this mode because we associate it with unpleasant feelings like hunger, cold, heat, or stress. We go to incredible lengths to live in a temperature- and humidity-controlled bubble with food constantly at hand. We develop countless devices and services to reduce physical effort. The remote control and the restaurant drive-through are just two examples.

Contrast this with our unicellular forebears. They were constantly switching between modes as they eked out an existence in a frequently hostile world. Threats and stressors were nearly constant, so they were well acquainted with the hunker down mode that is increasingly foreign to us. That's a shame because this mode has amazing health benefits. In times of plenty, we don't clean up old proteins. We stop cellular recycling. We tolerate leaky mitochondria. We neglect the imbalance in our immune systems. All of this happens because in the "grow" mode, we're mostly focused on the short term. Yes, we'll store excess energy away in the form of glycogen (liver and muscle) or fat (fat cells, eventually liver and muscle). But otherwise, our bodies are purely focused on the here and now—find a partner and reproduce. It turns out that exercise is part of the hunker down mode.

After all, if an organism is in a stressful environment because of nearby predators, changing conditions, or lack of food, what does it need to do? Two things—think and move. Both our mental and physical abilities are heightened during times of stress and privation. Exercise can help us mimic the hard times that bring out so many positive adaptations. In this chapter, we'll come to understand how and why exercise is so beneficial.

Benefits
....

The benefits of exercise are widespread and well documented. Study after study has demonstrated the many ways that exercise helps the body as well as the mind. It has been pointed out that humans are among the best endurance land animals. We were built to travel long distances for migration and hunting. Our bodies evolved to withstand prolonged periods of food deprivation as we explored new locales. From the perspective of evolution, our "normal" state is to be fit and trained. The sedentary condition in which most of us live should be considered an anomaly. When we listen to the evolutionary imperative to move, we gain access to the following benefits of exercise:

Controls weight and helps metabolic function. As we noted in an earlier chapter, perhaps the greatest health threat faced by people in developed countries is overnutrition. Among the many consequences of taking in too many calories are growth in fat stores (leading to obesity) and insulin resistance. Exercise helps burn off some of the excess energy stores and also helps maintain insulin sensitivity, especially in muscle. As a result, perhaps the greatest benefits from exercise come from its help in preventing us from gaining too much weight.

Enhances immunity. Regular cardiovascular exercise helps boost the immune system. It increases the number of natural killer (NK) immune cells out on patrol and looking for cells that have been infected by viruses or taken over by cancer.

Improves mood. Exercise increases circulation to the brain and helps alleviate anxiety and depression.

Boosts energy. People who engage in regular exercise tend to experience increased energy. Exercise promotes the recycling of old mitochondria and the production of new ones (mitochondrial biogenesis).

Promotes sleep. Exercise is the best treatment for insomnia and other sleep disorders. Exercise seems to reinforce circadian rhythms and the timely production of melatonin and other hormones linked to them.

Strengthens bones and joints. Contrary to what might be expected, exercise does not break down bones and joints with increased wear and tear. Instead, exercise serves to strengthen these vital components.

Improves sexual function. Sexual disorders such as erectile dysfunction can be greatly improved with exercise. This is in part due to the increased blood flow that comes with activity.

Can be fun and social. Exercise taps into the child-like desire to play games and track improvement. This is especially true in multi-person sports like tennis and basketball. Many other activities have a social component, including golf, skiing, and rock climbing.

Combats diseases and disorders. Exercise is well known to reduce the risk of just about all age-related diseases: diabetes, cancer, Alzheimer's, Parkinson's, heart disease, and many others.

People who are able to maintain a cardiovascular fitness level equivalent to gentle running enjoy a 50-percent reduction in all-cause mortality. And it's not just aerobic exercise that brings health benefits. High-intensity resistance training of as little as ten minutes a few times a week has demonstrated improvements in glucose control and type 2 diabetes.

As great as the benefits of exercise are, the downsides of inactivity are even more significant. Exercise requires coordinated activity by the whole body. The brain must activate to monitor the delicate operation of synchronously moving trillions of cells through space. The heart has to anticipate the coming demand and gear up to pump more blood. As exercise progresses, blood flow to non-essential areas is shut down. These areas include the gut, liver, kidneys, and uninvolved muscles.

It's easy to see what a powerful message exercise is to the body. But it's important to step back and look at the big picture. If an animal is exerting itself, it's because it's doing something it considers important—fleeing a predator, chasing prey, looking for a new food source, tracking a potential mate, etc. Exercise is a loud and clear message to biology—"This creature matters." So what message does prolonged inactivity send? "I have no purpose. I'm adding nothing of value. I'm not fleeing, chasing, hunting, building, or doing anything useful." Apologies for the stark characterization, but being inactive is like saying to the universe, "Take me."

Mechanisms

The benefits of exercise are well established. Aerobic fitness has been repeatedly demonstrated to reduce the risk of a wide variety of disorders including cardiovascular disease, cancer, depression, and type 2 diabetes. For many years, the mechanisms underlying these benefits were largely a mystery. In the last couple of decades, researchers have started to unravel how exercise produces these nearly miraculous results.

To understand how exercise provides its benefits, it's first necessary to appreciate the primary target of exercise—skeletal muscle. Skeletal muscle is all the muscle tissue around the body that is under conscious control. This does not include the heart or the smooth muscle used to constrict blood vessels or propagate intestinal contractions. At rest, skeletal muscle accounts for around 20 percent of cardiac output. That figure rises to 85 percent during exercise.

Muscle fibers come in two main flavors. Type I are "slow twitch" fibers optimized for endurance. They have large numbers of mitochondria and produce most of their ATP via oxidative respiration. These types of fibers burn a great deal of fat. Type II fibers are "fast twitch" and predominate in sprinters and other athletes who require short bursts of effort. They come in three varieties: types IIa, IIb, and IIx. Type IIb fibers have few mitochondria and rely primarily on glycolysis to produce energy. After using up their stored glycogen, these fibers turn to glycolysis and fermentation because they can't generate enough energy using mitochondria and ATP. These are the fibers that produce painful lactate when pushed too hard and too fast. Type IIa and IIx fibers are intermediate, between the other two types.

Fig. 103: Muscle fiber types

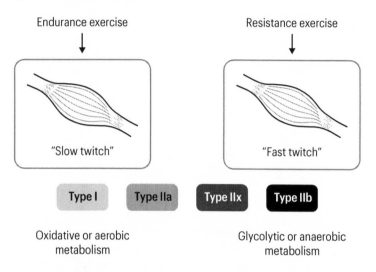

Aerobic exercise tends to increase the number and quality of mitochondria in type I muscle fibers. Resistance training increases the size and protein content of type II fibers and results in a small increase in mitochondria in those muscle fibers as well.

The best-known way to assess aerobic fitness is to measure VO₂max. This quantity calculates a person's maximum whole-body oxygen uptake during exercise. VO₂max is now recognized as a better predictor of mortality than any other risk factor or biomarker. As with other aspects of exercise, maximizing oxygen utilization involves optimizing multiple functions, including the amount of blood pumped by the heart, the resistance of the vascular bed to this pumping, the amount of skeletal muscle, the mitochondrial content of those muscle cells, the number of electron transport chains in those mitochondria, and so on. All of these parameters improve with exercise. As a result, exercise enhances just about every aspect of the body's operation.

Fig. 104: Exercised and trained skeletal muscle chart

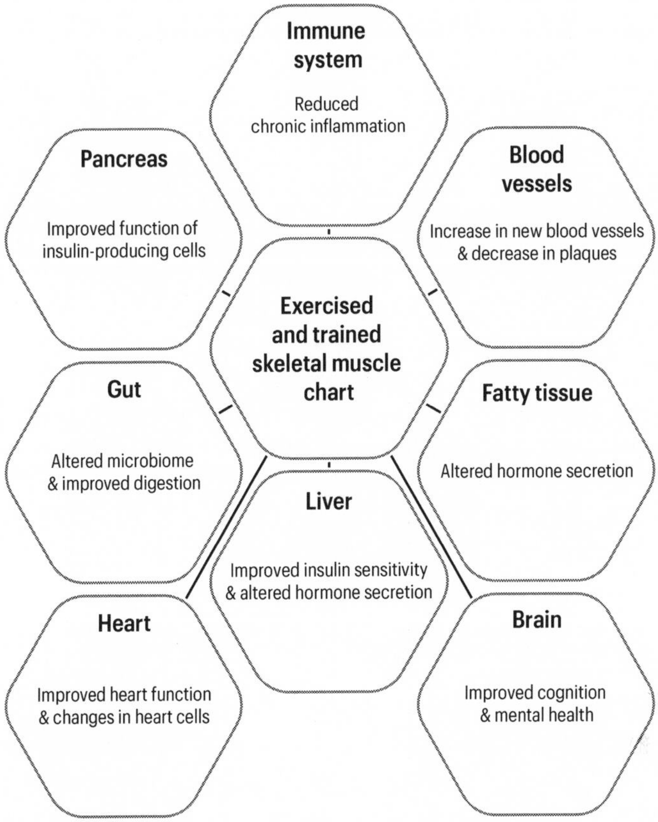

Let's dive down to the level of the muscle cell to see what's going on. As shown in figure 105, exercise begins with the contraction of a muscle fiber. The signal for that fiber to contract comes from a nerve that conducts an electrical signal to the neuromuscular junction (NMJ). This electrical pulse releases calcium from an internal storage depot that initiates the contraction. Calcium also activates certain proteins that travel to the nucleus and change the production (transcription) of various genes.

Fig. 105: Nucleus/NMJ/Mitochondria

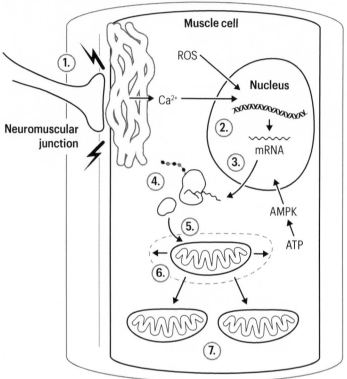

One important change that comes with regular exercise is the production of more mitochondria within muscle cells as well as other types throughout the body. This process is called mitochondrial biogenesis. We learned earlier that mitochondria are the descendants of bacteria and even today retain their own DNA. This DNA codes for thirteen different core components of the proteins in the electron transport chain—the bucket brigade that takes the electrons stripped from food and delivers them to oxygen, making ATP in the process. Recall that there are four large protein complexes that form the stations along this chain. For each complex, the core is coded by the mitochondrial DNA, while the rest of the proteins are encoded by DNA in the nucleus. The mitochondria need about a thousand proteins to operate, and the recipes for the vast preponderance of these are stored in the central library—the nuclear chromosomes. When new mitochondria are needed, often in response to repeated bouts of exercise, a signal is sent to the nucleus to ramp up production of these proteins. The process works like this:

1   The appropriate signal molecule travels to the nucleus, where it turns on the transcription of the genes needed for mitochondrial biogenesis.

2   The appropriate recipes are opened up and an mRNA copy of the recipe is made.

3   The mRNA is exported to the cytoplasm.

4   A ribosome forms around the mRNA and translates the recipe into the desired protein.

5   The protein is imported through a channel into the mitochondria throughout the cell.

6   The mitochondria form new electron transport chains, build new membranes, and grow larger.

7   A mitochondrion pinches off from one of the growing ones, and a new mitochondrion is born.

Signals for Mitochondrial Biogenesis

We've already talked about calcium's role as a signal to the nucleus by way of various calcium binding proteins. Two other important signals are reactive oxygen species (ROS) or free radicals and the protein AMPK, both of which we've discussed.

Recall that ROS are highly reactive forms of oxygen that arise when oxygen molecules in the mitochondria jump the line and steal electrons from complex I, II, or III instead of patiently waiting their turn at complex IV. This often generates a dangerous form of molecular oxygen called superoxide (O_2^-), an oxygen molecule with an extra unpaired electron. Superoxide is highly reactive and ready to form a bond with just about any other component in the cell—DNA, proteins, membrane lipids, you name it. This bonding can lead to mutations in DNA—especially the small amount of DNA that lives in the mitochondria. It can also damage proteins, including the sort of misfolding that can initiate harmful aggregates.

Luckily, the body makes two different enzymes that can stuff yet another electron (along with two protons) onto superoxide to yield the far more stable hydrogen peroxide (H_2O_2). This is only a temporary solution because hydrogen peroxide is itself a reactive oxygen species and can do the same sort of damage as superoxide. However, hydrogen peroxide reacts much more slowly. The downside is that this gives hydrogen peroxide time to diffuse throughout the cell before damaging components. It can even make it into the nucleus where it can damage the priceless nuclear DNA.

Although ROS might appear to be completely evil, they do serve a useful purpose. They act as signals to the rest of the cell that something stressful is going on. Low levels of ROS even indicate to the nucleus that it's time to consider building more mitochondria to share the load. For this reason, taking large doses of exogenous antioxidants can do more harm than good. Such antioxidants can blunt the sorts of beneficial adaptations that exercise prompts. Contrast this with the very low levels of hundreds of different kinds of antioxidants found in plants. These allowed low levels of ROS can continue to signal the need to make mitochondria while not building up to a dangerous concentration.

The third major signal for mitochondrial biogenesis comes in the form of a protein called *AMP kinase* or AMPK for short. We encountered this protein in the chapter on cell signaling and learned that it acts as a low-energy alert for the cell. Normally, there is far more ATP in the cell than AMP. Remember that ATP is analogous to a full AA battery, while AMP is that same battery after it has been run down. AMPK constantly monitors the ratio of AMP to ATP. When this ratio gets too high, the alert goes off. Think of the light on the instrument panel of your car that comes on when you run low on gasoline (or electricity). When AMPK is activated by a low energy balance in the cell, it starts to add phosphate groups to other proteins to turn them on or off. One of the proteins turned on by AMPK is called PGC-1a. In fact, we can think of all three signals—calcium, ROS, and AMPK—as leading to the activation of PGC-1a.

PGC-1a

PGC-1a is one of the handful of proteins you should become familiar with. It is called a transcriptional coactivator. Transcription factors are proteins that help decide which genes should be active or inactive at any given instant. In periods of extreme heat, for example, certain transcription factors travel to the nucleus and initiate the production of other proteins that help the cell deal with the high temperature. PGC-1a is not just a transcription factor but one of a handful of master regulators that coordinate the activity of many others. For our purposes, we can think of PGC-1a as the master switch for mitochondrial biogenesis. When it is activated, it initiates a whole protein cascade resulting in the production of brand-new mitochondria. As shown below, the results are twofold—more mitochondria and more electron transport chains per mitochondrion. PGC-1a decreases with age, but exercise can increase its effectiveness.

Fig. 106: PGC-1a

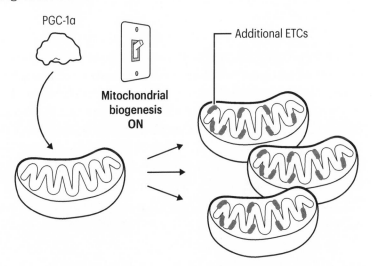

PGC-1α

Additional ETCs

Mitochondrial
biogenesis
ON

Stress Response

Our bodies have many systems to respond to both acute and chronic stress. We can think of stress as any threat to the sort of balance (homeostasis) we normally experience. Stressors can be physical (exercise, heat, cold), chemical (blood acid level or oxygen concentration), or psychological (fear, shame). The stress response has two main arms, the sympathetic nervous system and the hypothalamic-pituitary-adrenal (HPA) axis.

When the sympathetic nervous system is activated, we go into a mode often referred to as fight or flight. While activating the HPA axis takes minutes and can last hours or days, the sympathetic system turns on and off much more quickly. It is activated by a wide variety of stressors that trigger it to secrete the hormone adrenaline into the bloodstream. Adrenaline quickly increases heart rate, blood pressure, and other preparations for rapid response to threat or danger.

The HPA axis is the second part of the stress response. A part of the brain called the hypothalamus plays a central role. When stress is detected, the hypothalamus produces a hormone that travels a short distance to the pituitary gland at the base of the brain. The pituitary then secretes another hormone into the blood that travels to the adrenal glands on top of the kidneys. This hormone prompts the adrenals to secrete cortisol, the master hormone controlling the long-acting arm of stress response. Cortisol initiates a program designed to focus on the immediate cause of the stress—the sight of a predator lurking in the weeds, sounds of an invading clan, the taste of a poisonous plant, or the social stigma of embarrassment.

Regardless of the cause, the effect of cortisol on the body is largely the same—focus on the here and now at the expense of the future. Cortisol tells the liver to manufacture more glucose and pump it into the blood in anticipation of increased energy needs. It releases stores of white blood cells to be ready in case of injury. Cortisol heightens brain activity to help deal with whatever stress is at hand. However, the entire HPA axis is meant to be invoked for relatively short periods of time. As opposed to acute stress, which can actually be beneficial, chronic stress increases the probability of a whole range of diseases. Because the body is so focused on short-term survival, cortisol inhibits all manner of long-term maintenance projects. Blood is temporarily diverted away from the intestinal tract and kidneys in order to focus resources on muscle, heart, and brain. Synthesis of new white blood cells for the immune system is largely halted. Repairs to worn out cells in the skin and intestines are postponed. Why worry about routine maintenance when we're about to be attacked by a tiger?

In addition to the spikes triggered by acute threat, cortisol follows a regular pattern of rising immediately after waking in what is called the cortisol awakening response (CAR). Regular exercise decreases both the baseline cortisol levels as well as the peak levels experienced in response to new stressors. This may be because exercise optimizes the response of tissues to cortisol, so less is needed. The net effect is that fit people have less intense reactions to stress and require less cortisol for their response. The same is true for the sympathetic response. Fit people seem to deal more easily with stress, which allows their baseline levels of hormones like cortisol to be lower than in less fit people.

Inflammation

Once seen as an army garrisoned far away and summoned only under extreme threat, the immune system's place in our bodies is now known to be breathtakingly pervasive. In addition to protecting our cellular federation from dangerous interlopers like bacteria, the immune system keeps the peace within. We can think of our immune system as a whole set of federal services, passport control being just one of them. It also acts as a police force, dealing with rogue cells that attempt to subvert the union or simply lose their way. And we now know that the immune system is not confined to the blood or bone: Every cell in the body communicates in a vast network that flows through it. Each tissue has immune residents. The glial cells that support neurons in the brain are immune in nature. Peyer's patches in the gut, dendritic cells in the skin, Kupfer cells in the liver—the immune system is far from a remote citadel but rather is embedded deeply into the organism it serves.

As we've seen, the immune system takes a decidedly pro-inflammatory stance as we age. The trigger-finger myeloid branch comes to dominate the more discriminating lymphoid branch of T and B cells. This heightened baseline level of inflammation plays a role in every chronic disease, including diabetes, cancer, heart disease, Alzheimer's, and Parkinson's. One important effect of regular exercise on the body is decreased inflammation throughout. Inflammation can actually increase during the exercise session itself, but quickly returns to its normal level. Two important markers of inflammation, IL-6 and C-reactive protein (CRP), are lower in people who exercise regularly.

One of the ways exercise exerts its anti-inflammatory effect is by decreasing the amount of fat around the internal organs—the so-called *visceral fat*. It is increasingly clear that visceral fat is far more dangerous than fat stored in other locations. This "central" fat seems to produce high levels of inflammatory molecules called cytokines. Its strategic location close to the liver also seems to heighten its impact. Exercise helps reduce this visceral fat and decrease its tendency to spew out inflammatory cytokines.

A second source of inflammation that is ameliorated by exercise is the skeletal muscle itself. During exercise, muscles secrete IL-6 into the bloodstream to serve as a signal to the liver of the need to help support blood sugar levels. The muscles of physically active people are more sensitive to insulin and thus don't need as much support from the liver—and don't require as much of the IL-6 signal. Overall, the more active a person is, the lower their IL-6 levels, both at rest and when exercising. Given the strong association between chronic diseases and inflammation, this helps explain exercise's benefit across a wide range of disease conditions.

Neuroplasticity

This chapter has featured the repeated interplay between exercise and stress. Nowhere is stress more pernicious than in the brain. The brain areas most sensitive to stress—the hippocampus and prefrontal cortex—play critical roles in memory and executive function. We see the consequences of their damage in Alzheimer's.

Exercise has been demonstrated to enhance mood, improve cognitive function, and increase learning. These positive effects are due, at least in part, to a growth factor called brain-derived neurotrophic factor, or BDNF. BDNF helps dampen hunger and increase energy expenditure in certain brain areas. This growth factor is decreased by both chronic stress and inflammation. BDNF levels increase during exercise and remain elevated for about a third of regular exercisers. Exercise doesn't just prevent reduction in the hippocampus and prefrontal cortex—it actually increases brain volume in those areas. BDNF is an important part of the mechanism. In addition to promoting new neuron growth within the brain, it seems to help ward off insulin resistance.

A sedentary life is now so prevalent that it has become common to refer to exercise as having "health benefits," even though the exercise-trained state is the biologically normal condition. It is a lack of exercise that is abnormal and carries health risks. In fact, overall fitness all by itself (independent of smoking or other risks) is an excellent predictor of a person's probability of dying or suffering from disease in the next few years.

Exercise as a Stress Buffer

Taken together, it's evident that the effects of exercise combine to increase a person's ability to withstand both acute and chronic stress. Better fitness of the heart, lungs, muscles, brain, and other organs means that less cortisol is needed during stress events. This helps avoid some of the negative consequences of cortisol

and other stress-related hormones. Interestingly, the stress toler-
ance induced by exercise applies even to psychological stressors.
Going back to our trampoline analogy, exercise strengthens not
only the bed of the trampoline but also the springs and other components, and it helps the trampoline withstand the sudden and
surprising loads that life is so apt to throw at us. The benefits are
far-ranging, enhancing the nervous system, metabolic processing, the inflammatory environment, and the endocrine system, as
well as improving the muscle tissue itself. And you don't need to
engage in vigorous exercise to realize these benefits. Just walking
is enough to increase your fitness, especially if you do it for at least
fifteen minutes.

References and Further Reading

SR Gilani et al (2019). The effects of aerobic exercise training on
mental health and self-esteem of type 2 diabetes mellitus patients.
Health Psychology Research, 2019(7): 6576.

SL McGee and M Hargreaves (2020). Exercise adaptations: molecular
mechanisms and potential targets for therapeutic benefit. *Nature
Reviews Endocrinology*, 16: 495–504.

SC Moore et al (2016). Association of leisure-time physical activity
with risk of 26 types of cancer in 1.44 million adults. *Journal of the
American Medical Association, Internal Medicine*, 2016(176): 816.

DM Peterson (2013). Overview of the benefits and risks of exercise. Online:
uptodate.com/contents/the-benefits-and-risks-of-aerobic-exercise.

MN Silverman and PA Deuster (2014). Biological mechanisms underlying the role of physical fitness in health and resilience. *Interface
Focus*, 4: 20140040.

RM Simon et al (2015). The association of exercise with both erectile and sexual function in black and white men. *Journal of Sexual
Medicine*, 12: 1202.

J Vina, F Sanchis-Gomar, V Martinez-Bello, and MC Gomez-Cabrera
(2012). Exercise acts as a drug; the pharmacological benefits of
exercise. *British Journal of Pharmacology*, 167(1): 1–12.

14

Intermittent Fasting

THE MOST fundamental stress facing any organism is starvation. Whether we are a bacterium cork-screwing through the lonely ocean or a human staggering through a remote forest, our response is the same. All our faculties are trained on the objective at hand—obtaining enough energy to survive. Nothing focuses the mind like the prospect of starvation. Concerns that seemed so critical before now fade to irrelevance. The only thing that matters is finding the next meal.

Before attempting to understand how periodically letting our bodies get a little hungry might be beneficial, let's step back and take a look at the big picture. Imagine we're building a robot. We want our robot to be able to explore a remote planet like Mars. When the sun is out, our robot will deploy solar panels and generate electricity for immediate use. However, there may be long periods when no sunlight is available. How would we approach this problem as engineers?

Well, the first step is to have some energy available for the periods when the sun first goes away. Batteries are the obvious choice. We'll build batteries in our robot that will provide enough energy for most dark periods—so, once fully charged by the sun, our robot will be able to function for about twelve hours before running low on energy. What then? Do we just park the robot and wait for the sun to come back out? That might work, but what if we enter a prolonged sunless period? If the robot spends too much time without power, its batteries might freeze and become inoperable. Clearly, we need a backup plan.

How about if we add a little fuel cell to the robot and a tank that holds hydrogen gas? When the batteries start to run low, we can turn on the fuel cell and let it provide energy until we regain the sun. Let's assume that our robot can run off a full tank of hydrogen for up to a month. It's almost certain that the sun will come back out to recharge the batteries within that period of time. This scheme works for the first prolonged dark period, but what happens once the robot has used up all its stored hydrogen? What if we could replenish the hydrogen using sunlight? This way, when we do have sunshine, we can first recharge the batteries and then use any remaining sunlight to generate enough hydrogen to refill the tank.

But what will the sunlight act on to generate hydrogen? The most obvious candidate is water. Let's add a special compartment to our robot that can hold a bucketful of water or ice. As the robot travels around, whenever it encounters water or ice, it can scoop some up and throw it in this special compartment. It can leverage this source of H_2O along with electricity generated by solar panels to split the water into hydrogen and oxygen. We'll let the oxygen escape and push the hydrogen gas into the fuel cell. This leaves us with an elegant three-level energy system:

1   Solar panels to generate electricity when the sun shines. This electricity can be used to power all the robot's systems.

2 Batteries that can be charged by the solar panels to provide power when the sun isn't around. This is the first source we'll tap into when it becomes dark.

3 A tank of hydrogen gas that can flow through a fuel cell to generate electricity. We'll need to tap into this third energy source only during prolonged periods without sunshine.

Of course, no system is perfect. If the planet our robot is exploring suddenly experiences months of darkness, our robot will run down and freeze. However, we've designed it so that such an unfortunate event is unlikely to occur.

It may surprise you to learn that mammals (and other animals) have a similar energy architecture that consists of the following sources:

1 The food in your stomach. Right after a meal, this generally provides enough energy to power your body for about eight hours.

2 The glycogen in your liver. Glycogen is the storage form for glucose. The liver builds it up when you eat and then uses it to maintain blood sugar levels during fasting periods. This will provide roughly four more hours of energy.

3 The fat around your middle. Fat is an even denser energy storage form than glycogen. The average person has enough fat to provide all their energy needs for at least a month. Your body begins to break down this fat and use it for energy about twelve hours after your last meal. This process escalates, and by sixteen hours after your last meal you are breaking down and burning fat like crazy.

Our bodies perform best when we cycle through all three of these energy sources on a regular basis, as we'll soon see. Unfortunately, our modern lifestyle usually means that we seldom touch third-level storage and our fat reserves just continue to grow.

Permanent Caloric Restriction

In the middle of the twentieth century, scientists studying the effects of starvation were stunned to make a paradoxical observation. Animals fed reduced amounts lived longer and seemed to avoid many of the diseases associated with aging as long as the caloric reduction didn't tip them into overt malnutrition. That is, eating less while still consuming enough increased both lifespan and health span.

How in the world could this be? We all know that our bodies—and our cells—require nutrition and energy to survive. How could supplying them with less be beneficial? As mentioned in the chapter on autophagy, these experiments were revived in the last twenty years and multiple groups have now demonstrated the effectiveness of restricting calories by one third or so in improving health and even increasing lifespan. And they showed this in one of our closest relatives, rhesus monkeys. There is even an organization called CR Society International that consists of thousands of people attempting to live on approximately twelve hundred calories per day—permanently. Early results show adherents to be healthier than their societal counterparts who eat normally. Study after study has confirmed that permanently restricting calories in any animal, including humans, will decrease the burden of disease and increase lifespan.

Intermittent Caloric Restriction

Unfortunately, very few of us have the discipline to severely restrict our food intake for the rest of our lives. That's why many people around the world were excited when word first started to leak out roughly ten years ago that the same benefits could be enjoyed with sporadic caloric restriction—what has come to be known as intermittent fasting. If anything, intermittent fasting appears to be even more effective. Exactly what is intermittent fasting and why is it different from or better than other "diets"?

First, it's important to realize that intermittent fasting is not a diet. In most studies, subjects continue with their normal meals but just limit the times during which they can consume that food and drink. For example, one common approach is to pack all calories into an eight-hour window, perhaps from 10 a.m. until 6 p.m. every day. This means no calories from 6 p.m. until 10 a.m. the next morning—a sixteen-hour fast. Of course, there's nothing magical about sixteen hours, although the introduction to this chapter may help explain why this duration is a good place to start.

Researchers and practitioners have experimented with a wide variety of approaches. Some studies allow participants to eat normally one day and then limit themselves to fewer than five hundred calories the next day. Some fasting adherents eat only one meal a day (OMAD) and try to consume their entire day's worth of calories in a window of four hours or less. Other fasting programs allow participants to eat normally for five days and then fast for two (5:2 plan). The main objective of all these approaches is to build into each week a period of time during which the body takes a break from food.

Mechanism

When the body runs low on energy, alarm bells start going off. The alpha cells in the pancreas sense low blood glucose levels and secrete a hormone called glucagon, a sort of mirror image of insulin. Whereas insulin tells cells throughout the body that times are good and they should grow and sock away energy for a rainy day, glucagon sends cells a warning: glucose levels are low, and they need to take the appropriate steps to protect themselves. In particular, the liver picks up on this glucagon message and does what it can to help. First, it breaks down any glycogen (stored glucose) it has and pumps this into the blood to maintain blood sugar levels. Once that glycogen is gone, the liver starts to manufacture new glucose molecules in a process called gluconeogenesis.

Fig. 107: Ketones in system

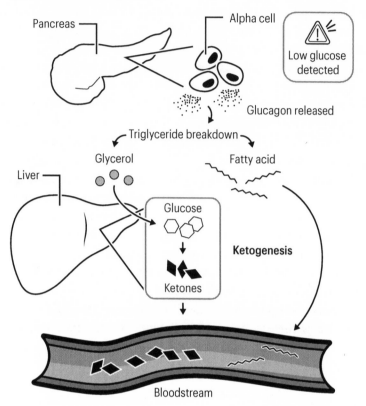

Glucagon also signals fat cells to break down their stored triglycerides. Recall that a triglyceride consists of one molecule of glycerol and three molecules of fatty acids. When fat cells break down their stored triglycerides, the fatty acids flow into the blood, where they can be absorbed by various cells around the body—especially the heart and muscle—and oxidized in the mitochondria of those cells to generate ATP. The glycerol from these triglycerides travels through the blood to the liver, where it can be used to make new glucose molecules. The most common fatty acid in mammals, palmitate, is a string of sixteen carbon atoms. These long molecules don't pass well into the brain, so the liver will absorb some of them, chop them into fragments just two or three carbons long, called ketones, and then pump the ketones back into the blood. The brain and other organs are able to absorb these ketones and then burn them in their mitochondria to produce ATP. This process is illustrated below.

So far, we've been looking at the effects of low energy at the whole-body level. Now let's dive down to the level of the cell. As mentioned previously, the cell's low-energy sensor is AMP kinase—AMPK. This protein is constantly monitoring the ratio between ATP (full battery) and AMP (empty battery). When that ratio falls below a certain level, AMPK is activated and springs into action. AMPK's job is to go around the cell and paste phosphate molecules (PO_4^{3-}) onto other proteins. This has the effect of turning some proteins off and others on, allowing AMPK to have a far-reaching impact throughout the cell. We covered some of the details regarding AMPK back in the chapter on cell signaling. AMPK activates the SIRT1 protein in the cytoplasm as well as its cousin, SIRT3, in the mitochondria. These two sirtuins are stress detectors and set a number of actions in motion within the cell. Because of the low-energy alert they've received from AMPK, they "know" that the cell needs to take rapid steps to generate more energy. These include:

1 Pushing current mitochondria to burn more fat and generate more ATP.

2 Working with PGC-1a to initiate the construction of new mitochondria.

3 Telling the mitochondria to generate antioxidants.

4 Telling the nucleus to transcribe the genes coding for antioxidants and chaperones.

5 Increasing the rate of autophagy to recycle old components to provide new building blocks.

We talked a bit about senescent cells in an earlier chapter. There's some evidence that fasting helps push senescent cells over the edge, leading to their death and replacement. Intermittent fasting increases insulin sensitivity so the baseline level of insulin in the body drops. So does inflammation. Intermittent fasting tends to reduce the level of inflammation throughout the body.

Phases of Fasting

We can divide a fasting period into the following phases.

Hours 0–8. For about eight hours after a meal, the body is able to meet its energetic needs by breaking down the food from that meal.

Hours 8–12. Once the stomach is empty and blood sugar starts to drop, the liver begins to break down glycogen, its long chain of stored glucose molecules. This can maintain blood sugar levels for around four hours.

Hours 12–16. Once the liver's glycogen is gone, blood sugar levels begin to slowly dip. Signals go out to fat cells to start to break down their reserves and release fatty acids into the blood. Cells in the heart, muscles, kidneys, and other tissues begin to burn these fatty acids in their mitochondria to produce energy. The liver absorbs some of the fatty acids and chops them up into smaller ketones that it pushes back into the bloodstream. These ketones are primarily used by the brain. For the first few hours, this process occurs at a low level and is called light nutritional ketosis.

Hours 16–24. By roughly sixteen hours into a fast, the breakdown and burning of fat accelerates. So, too, does the rate at which the liver produces ketones for the brain. The body has entered full ketosis.

Hours 24–48. Autophagy has been increasing during the entire fast, but by twenty-four hours it is in full swing. Cells throughout the body are breaking down old proteins, some of which might have unfolded and formed small aggregates. Old mitochondria are being recycled as well. As time goes on, cells start to recycle more aggressively. The long fasting period spurs the production of growth hormone. Forty-eight hours into a fast, growth hormone levels have increased fivefold.

Hours 48–72. With longer fasts, the body becomes more and more sensitive to insulin, so less is needed. By fifty-four hours into a fast, insulin has hit a minimum.

Hours 72–96. After three days of fasting, the body begins to break down old immune (white) cells and the bone marrow replaces them with new ones. This has the beneficial effect of restoring the youthful 50/50 balance of myeloid (inflammatory) and lymphoid cells, thereby reducing inflammation.

Fasts longer than ninety-six hours are dangerous and should be done only under the supervision of a physician. After a few days, the body has no choice but to start breaking down the protein in muscle to supply its energy needs. There are ways to minimize this effect, and long-duration therapeutic fasts are utilized in clinical settings in Germany and other countries. However, so-called dry fasting is downright stupid. Be sure to maintain fluid intake during all fasts.

Please note that the timeline described above is approximate. People following a low-carb diet may enter ketosis much more rapidly than someone fueling up on beer and pizza. Some people may not start heavy ketosis until close to twenty-four hours after their last meal. The best approach is to use a ketone testing device to determine your own body's response to fasting.

Benefits

Even though the scientific basis of intermittent fasting has been unraveled only in the last twenty years, humans have recognized the benefits since antiquity. Benjamin Franklin said, "To lengthen thy life, lessen thy meals." Fasting is part of many religious traditions.

The evidence on intermittent fasting is clear. The practice improves the function of every organ and functional system in the body. We've already talked about autophagy, the cell's primary recycling process. Undoubtedly many of the benefits of intermittent fasting derive from its ability to significantly drive up the rate of this recycling. Autophagy takes place all the time, but usually at a very low level. In the modern world, by eating nearly around the clock, we tend to keep autophagy mostly shut off. Intermittent fasting is an easy and powerful way to initiate this beneficial process. Figure 108 in the next section illustrates some of the many benefits that can result from simply taking a short break from nutrition.

Among the most interesting of these benefits are the effects on the brain. The stomach produces a small hormone called ghrelin, which is the source of the hunger pangs we feel when our stomachs are empty. It turns out that this hormone, along with other signals during fasting, increases the production of a protein called brain-derived neurotrophic factor (BDNF). BDNF spurs the creation of new neurons that are critical for learning and memory.

Overall, intermittent fasting is one of the most powerful ways to activate hormesis. Human ancestors frequently had to survive long periods without food as they migrated around the world. Nature endowed them with the ability to withstand these long periods of nutritional deprivation. Hunger increases alertness and brain function to help with the mental work required to identify food sources. It initiates the breakdown of fat deposits around the body, especially the dangerous ectopic fat that has been stored in the wrong places (outside fat cells) during times of excess. Like exercise and the other stressors covered in this book, intermittent fasting can help create a reserve against other stressors and thus improve overall health.

Medical Considerations

Intermittent fasting is beginning to be featured in various clinical trials and medical case studies. Results are just starting to trickle in, but the early returns are very promising. Numerous studies report being able to reverse type 2 diabetes and even take patients off insulin. Intermittent fasting appears to have benefits in all organ systems throughout the body.

Fig. 108: IF considerations

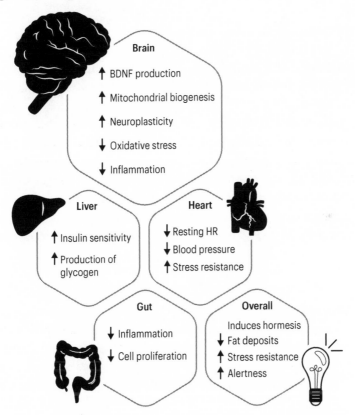

Various clinical trials are exploring the use of intermittent fasting in treating cancer. In one study, patients underwent a fast for two days before receiving chemotherapy. The results were decreased side effects and improved killing of cancer cells. Just searching the database of trials at clinicaltrials.gov for the term *fasting* yields thousands of results.

WE'VE NOW covered three simple ways to exploit slight stressors to provoke an adaptive response: eating healthy plants, exercising, and intermittent fasting. As we'll see in the next chapter, we have even more options at our disposal.

References and Further Reading

SD Anton et al (2017). Flipping the metabolic switch: understanding and applying the health benefits of fasting. *Obesity*, 26(2): 254–68.

R De Cabo and MP Mattson (2019). Effects of intermittent fasting on health, aging, and disease. *New England Journal of Medicine*, 381: 2541–51.

S Furmli, R Elmasry, M Ramos, J Fung (2018). Therapeutic use of intermittent fasting for people with type 2 diabetes as an alternative to insulin. *BMJ Case Reports*. October 9, 2018.

VD Longo and MP Mattson (2014). Fasting: molecular mechanisms and clinical applications. *Cell Metabolism*, 19(2): 181–92.

15

Other Stressors

"Do not be afraid of discomfort. If you can't put yourself
in a situation where you are uncomfortable, then you
will never grow. You will never change. You'll never learn."

JASON REYNOLDS

S O FAR, we've examined three major stressors—exercise, plant-based chemicals, and fasting—and how our bodies and cells respond to them. In this chapter, we'll briefly cover some other stressors that evoke the same type of hormetic response. As with the other factors we've looked at, these stressors also cause the body to take countermeasures to adapt to the stress and get ready for worse to come. As long as these stressors are relatively mild and repeated, they will generate the same "what doesn't kill you makes you stronger" reaction, right down to the level of the individual cell.

Cold

....

As human ancestors walked out of Africa to spread across the globe, they encountered a variety of challenging environmental conditions. Cold was certainly one of these. It makes sense that the body would sense the threat posed by low temperatures and take adaptive measures. In fact, we're finding that exposure to cold is a powerful force to induce hormesis. You might have heard of Wim Hof—the Dutch "Iceman" who gained fame with his exploits, which involved swimming under the ice in frozen rivers, withstanding prolonged exposure to freezing temperatures, and holding his breath for a record-setting length of time. We're now starting to understand the mechanism underlying the benefits that come from exposure to cold.

It has recently been discovered that fat cells (adipocytes) come in two main colors—white and brown. White fat cells are the ones we're most familiar with. They are mostly simple storage depots for triglycerides, although we're learning that they secrete inflammatory cytokines and other hormones. This is especially true of the white fat cells that coat the internal organs in our abdomen—so-called visceral fat.

We recently discovered that some animals have a different kind of adipocyte called brown fat. Both white and brown fat cells have mitochondria and use the electron transport chain to generate ATP. Recall that ATP is made when protons flow through a little machine called ATP synthase in the inner mitochondrial membrane. The electron transport chain has four stops—complexes I through IV—and then ATP synthase. It turns out that brown fat cells have pores called uncoupling proteins that short-circuit the electron transport chain and allow protons to flow directly back into the matrix of the mitochondria without passing through the ATP synthase turbine.

Why would this be? Instead of using the potential energy of the proton gradient to squeeze a phosphate onto an ADP molecule, brown fat cells use this energy to generate heat under cold conditions. It would be like tearing down our waterwheel (in an earlier chapter). Instead of turning the blades on the wheel and thereby generating the force required to turn a grinding stone (or turn a turbine to produce electricity), the water simply falls onto the ground. The potential energy is turned into kinetic energy, which heats up the ground. Brown fat cells are taking the same approach—letting the protons back into the matrix in a way that generates heat instead of ATP.

Fig. 109: Adipocytes

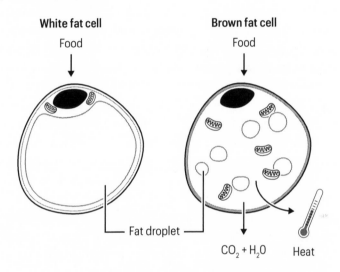

As you might imagine, brown fat has a higher metabolic rate than white fat. Because of this, it is protective against diabetes and other diseases. It was thought that humans didn't have brown fat, until it was found in infants. Subsequently, researchers discovered that being exposed to cold temperatures encourages white fat cells in human adults to become brown, or at least beige. Such fat cells burn more energy than normal white fat cells and thus are beneficial to health. This sort of adaptation can happen in the reverse. That is, people from cold climates who might have built up reserves of brown fat can lose them if they move somewhere warm. When they return to their native environment, they might report feeling colder because they have less brown fat to generate heat. Snowbirds beware.

Repeated mild cold exposure has many other positive effects. Overall, it slows down protein metabolism and decreases the rate of cell division (proliferation). These effects are obviously contrary to the needs of cancer cells. Another benefit of repeated exposure to cold is increased insulin sensitivity. This can partly be explained by the development of brown adipose tissue, as discussed above, but there seem to be additional mechanisms. One study reported that ten days of exposure to mild cold (15°C or 59°F) was sufficient to increase insulin sensitivity by almost half. Skeletal muscle fibers began to display more GLUT4 glucose transporters, allowing them to absorb more glucose for the blood at a lower level of insulin signaling.

Fig. 110: Cryotherapy

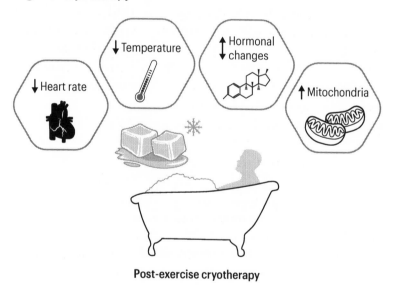

Post-exercise cryotherapy

Cryotherapy is a form of cold therapy that has become popular in recent years. Generally, athletes enter a chamber after exercise that allows the entry of super-cooled nitrogen gas, which quickly lowers the temperature of the outside of their body. These treatments usually last two to three minutes at temperatures as low as -200°F (-130°C). The purported benefits are decreased soreness after an intense workout and lower levels of inflammation. Cryotherapy can also encourage the formation of new mitochondria (mitochondrial biogenesis) via our old friend PGC-1a, and reduce the heart rate. This is why tennis players take an ice bath after a match.

Other studies have demonstrated that cold-water swimmers have reduced levels of inflammation and a lower baseline of stress hormones such as cortisol. As with other mild stressors that activate hormesis, the overall effect seems to be to increase resilience to stress of any form.

Heat

As with cold, heat is a stressor *Homo sapiens* have had to deal with from the time they appeared on this planet. Repeated exposure to heat can lower blood pressure, increase blood volume, and decrease core body temperature. Several studies have examined the effects of frequent sauna use in Finland. In Finnish saunas, hot and dry air circulates at 80 to 100°C (176 to 212°F). The average Finn takes a sauna twice a week for anywhere between five and thirty minutes.

Heat causes a number of physiological adaptations. In particular, more blood flows to the skin in an effort to dissipate the excess temperature, which reduces blood pressure and increases heart rate. Blood flow increases to muscles and bone marrow. Most impressively, regular sauna use significantly decreases risk of heart disease, Alzheimer's, and dementia in general. We're not talking small risk reductions here, but on the order of 65 percent for those taking saunas four or more times a week.

How does taking regular saunas induce these near miraculous effects? It seems that sauna sessions mimic exercise by increasing blood flow. This increased blood flow is a mild stress on the vascular system and forces the cells lining blood vessels (endothelial cells) to up their game by strengthening cell-to-cell connections and other adaptations. The heart and circulatory system increase baseline fitness, thus providing more of a reserve to respond to other stressors. Saunas also cause a mild oxidative stress, causing cells to produce more antioxidant enzymes and heat shock proteins to help keep proteins properly folded. Other studies have found that mice exposed to regular mild heat stress have far fewer of the plaques and tangles associated with Alzheimer's in their brains.

Low Oxygen (High Altitude)

Why do people living in Boulder, Colorado, have the lowest rate of obesity in the United States—one third the national average? It appears that people around the world who live at higher elevations have lower rates of obesity, diabetes, and other diseases—even when controlled for other differences. One study showed that people living at five thousand feet (fifteen hundred meters) above sea level (including residents of Boulder) lived two to three years longer than their lowland counterparts.

Breathing "thin" air, with less than the normal amount of oxygen, is a mild stress like so many of the others covered in this book. One important effect of hypoxia on the body is a reduced rate of protein synthesis. Protein stress increases with age. Misfolded proteins and their aggregates increase rapidly and play important, perhaps causative, roles in diseases like Alzheimer's, Parkinson's, ALS, and even type 2 diabetes. Recall that the protein mTOR is a key sensor for amino acid levels and overall nutritional status within the cell. When mTOR is active, protein synthesis accelerates and autophagy (recycling) shuts down. This is appropriate when an organism is young and growing, but mTOR activity continues at a high level as we age. Hypoxia activates a transcription factor that inhibits the protein mTOR, which some researchers view as contributing to the aging process. The cell interprets the hypoxia as a stress that must be dealt with and diverts energy into addressing it and preparing for even worse.

Radiation

X-rays, ultraviolet rays, and ionizing radiation can all penetrate tissues and damage DNA. From the 1950s on, the conventional view was that there was no safe level of radiation and all radiation was bad. This is the so-called linear no-threshold (LNT) model. Over time, however, it has become clear that even radiation can invoke a beneficial hormetic response at low doses. Figure 111 comes from a recent study looking at Japanese survivors of the Hiroshima atomic bomb. The rates of leukemia observed were far lower than expected until exposure reached high levels. In fact, exposure of up to about fifty rem actually *reduced* rates of leukemia compared to people with no exposure. Once again, we see the U-shaped response curve of hormesis.

Figure 111: Leukemia graph

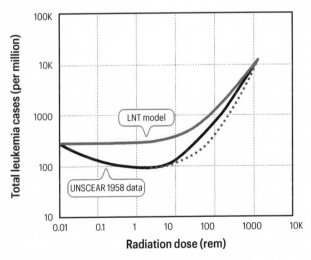

Adapted from Cuttler (2015).

Some clinicians contend that low-dose radiation can be used clinically to treat Alzheimer's, Parkinson's, and other disorders. The precise mechanism through which such radiation might be helpful is not known, but speculation centers on improvements in immune function and possible clearing of senescent cells. Other groups believe that repeated electromagnetic field stimulation can reduce the accumulation of beta amyloid in people with Alzheimer's.

Shortly after radiation was discovered at the beginning of the twentieth century, it was touted for its salutary benefits. Health retreats opened at hot springs and other sites of elevated background radiation. These retreats quickly closed once the cancer-causing effects of high-dose radiation were discovered, but exposure to low-dose radiation is now back in vogue. The Radium Palace is a popular facility in the Czech Republic that uses low-dose radiation to treat a range of common disorders.

One proposed mechanism for radiation hormesis is the conversion of small amounts of water in the cell to hydrogen peroxide. This might be enough to activate the NRF2 pathway—similar to the effects of sulforaphanes in broccoli (and especially broccoli sprouts).

Even More

If you're getting the impression that just about any stressor can be beneficial, you're probably right. As long as cells can recognize the stress, they can mount a response. If the stress is at a low enough level, the response will probably go beyond the initial stress and thus confer protection, even against completely different kinds of stress. Low levels of hydrogen sulfide in mineral baths are reported to have a beneficial hormetic effect. Even low levels of psychological stress can trigger hormesis! Clearly, we and our cells are built to handle a wide variety of stressors. Change, including challenging change, keeps us alive. It's stasis that kills us.

References and Further Reading

A Bartelt and J Heeren (2014). Adipose tissue browning and metabolic health. *National Review of Endocrinology*, 10: 24–36.

J Cuttler and JS Welsh (2015). Leukemia and ionizing radiation revisited. *Journal of Leukemia*, 3(4).

HA Daanen and WD Van Marken Lichtenbelt (2016). Human whole body cold adaptation. *Temperature*, 3(1): 104–18.

S Dattilo et al (2015). Heat shock proteins and hormesis in the diagnosis and treatment of neurodegenerative diseases. *Immunity and Ageing*, 12(20).

M Ezzati et al (2012). Altitude, life expectancy and mortality from ischaemic heart disease, stroke, COPD and cancers: national population-based analysis of US counties. *Journal of Epidemiology and Community Health*, 66(7): 1–8.

MJW Hanssen et al (2015). Short-term cold acclimation improves insulin sensitivity in patients with type 2 diabetes mellitus. *Nature Medicine*, 21(8): 863–65.

I Heinonen et al (2018). Effects of heat and cold on health, with special reference to Finnish sauna bathing. *American Journal of Physiology-Regulatory, Integrative, and Comparative Physiology*, 314: R629–R638.

NR Mane et al (2018). Mild heat stress induces hormetic effects in protecting the primary culture of mouse prefrontal cerebrocortical neurons from neuropathological alterations. *IBRO Reports*, 5: 110–15.

JM Peake (2017). Cryotherapy: are we freezing the benefits of exercise? *Temperature*, 4(3): 211–13.

H Sies and LE Feinendegen (2017). Radiation hormesis: the link to nanomolar hydrogen peroxide. *Antioxidants and Redox Signaling*, 27(9): 596–98.

MJ Tipton et al (2017). Cold water immersion: kill or cure? *Experimental Physiology*, 102(11): 1335–55.

16

Recommendations for a Long, Healthy Life

"The science of being healthy is well-known. It is not
esoteric. There are no magic bullets. If you want to live
a long life, we've known the answers for more than
a hundred years. It's a wide-ranging diet with as much
fruit and veg as you can stuff into yourself, and plenty
of exercise. It doesn't even matter what kind of exercise."

ALICE ROBERTS

S O FAR in this book, we've spent a great deal of time covering
how life developed. We've learned that our cells still have a
lot in common with the ancient organisms that pioneered
life on our planet nearly four billion years ago. In fact, we've seen
that ancient survival programs are still intact in the trillions of
cells that comprise our own bodies. Hopefully it has become
obvious that one of the easiest ways to improve our own health
is to activate the hormesis program in every possible way. We
can do this by moving outside our modern comforts and forcing ourselves to endure relatively minor stressors in the form of

mild plant-based toxins, exercise, hunger, cold, and heat, among others.

In this chapter, we'll use all this information to make a list of recommendations for living a long and healthy life. Note that these recommendations all have a strong scientific basis and are not merely opinions or generalizations of anecdotal studies. The measures described in this chapter are designed to do the following:

* Avoid the many dangers and adverse effects of overnutrition

* Improve insulin sensitivity

* Decrease the encouragement that high insulin levels provide to cancer cells

* Reduce the overall level of inflammation throughout the body

* Force cells to recycle old components to avoid protein aggregates

* Regularly destroy poorly functioning mitochondria and replace them with new ones

* Increase overall ability to withstand acute stressors

* Restore a healthy 50/50 immune balance across myeloid and lymphoid lines

* Reduce oxidative stress and the damage it can wreak on DNA, proteins, and lipids

* Optimize the stress response system so that less cortisol is needed at baseline

* Maintain muscle mass

* Improve cardiovascular fitness

* Strengthen muscles critical in day-to-day life

* Enhance sleep quality and duration

As noted with exercise, if you could get these results from a medication, you would probably pay a fortune for it. The recommendations outlined below will take a bit more work than popping a pill, but they are inexpensive and will do more for your health than any medication ever could.

Eat at least five servings of healthy fruits, vegetables, seeds, nuts, and oils each day. The various phytochemicals in these plants have been demonstrated to improve human health in a variety of ways. In addition to functioning as antioxidants, many of them also invoke hormesis by acting as minor mitochondrial toxins. This causes mitochondria to signal to the nucleus that they are under stress and to come to the rescue with more chaperones to assist in proper folding and various classes of antioxidants. The fiber and phytochemicals in plants also help promote a healthy gut microbiome.

Note that you don't have to become vegan or vegetarian to follow this recommendation. Just get a variety of healthy plant-based products into your mouth every day in addition to the beer, pizza, hamburgers, and other foods we all love. Smoothies are one easy way to achieve this objective. Here's the author's favorite recipe: blend together one banana, a cup of strawberries, a few blackberries or blueberries, a handful or two of baby spinach, two tablespoons of ground flaxseed, a couple tablespoons of unsweetened Greek yogurt, a half cup of unfiltered apple juice, and a handful of ice cubes. This alone will deliver your five servings of healthy plants. Enjoy your smoothie along with some bread that you soak in extra virgin olive oil for an even healthier snack.

Adopt intermittent fasting on a regular basis. Try to force all your calories into a relatively narrow window—ideally from four to ten hours most days of the week. That is, attempt to fast between fourteen and twenty hours each day. Most importantly, don't take in any calories once you've finished dinner until your first meal the next day. It's fine to experiment with longer fasts up to twenty-four hours or so, but you don't need to perform multi-day fasts to get the benefit. After roughly sixteen hours of fasting, your body will be in full fat-burning mode. Over time, this will improve your insulin sensitivity and reduce your baseline insulin levels. It's true that some fasting adherents prefer to do one or two longer fasts (e.g., twenty-four hours) each week. However, much psychological research suggests that most people have more success developing daily habits.

Reduce inflammation. Inflammation plays a large part in every chronic disease and has been featured throughout this book. Inflammation can be reduced by losing excess weight, improving cardiovascular fitness, and avoiding sugar and other pro-inflammatory foods. Just about all of the recommendations in this chapter will serve to reduce inflammation.

Subject yourself to discomfort. Take a walk in freezing weather. Put on a coat and go out in the rain and snow. Take a periodic sauna or ice bath. If you have the opportunity, spend time in the mountains and hike at elevation. Don't be fixated on always being comfortable. As we've seen in previous chapters, all these forms of discomfort can trigger an adaptive response that makes you stronger and healthier. Remember that the molecule responsible for hunger pangs (ghrelin) causes your body to produce more BDNF to help fight against the brain atrophy that often accompanies aging.

Work on your cardiovascular fitness. What's good for your heart is good for your brain and other organs. Cardiovascular fitness brings a wide variety of benefits—lower blood pressure, increased blood flow to vital organs, increased ability to handle stress, improved mood, better sexual function, and many more. There's really no excuse. You don't need to belong to a gym, and you don't have to buy any expensive exercise equipment. Walking is the simplest form of cardiovascular exercise, but there are countless ways to improve your aerobic fitness. Just stepping up on a small stool or block while watching TV will do it. Jumping rope, walking stairs, jogging—all will increase your cardiovascular fitness if you do them regularly. Make it a point to get at least thirty minutes of cardiovascular exercise every day.

Develop your core muscles. Doctors, physical therapists, and other experts have come to appreciate the importance of core stabilizer muscles, especially as people age. These are the muscles that help us get out of a chair and prevent a fall. We use them in all sorts of day-to-day activities. If we let these muscles go, we're sure to suffer a variety of problems, including back pain. One easy exercise you can use to develop your core is the old-fashioned plank. Stretch well before attempting, but see how long you can hold this position. Work up to the point that you can hold this plank for at least sixty seconds, and do that every day. There are many variations you can play with as you get better.

Fig. 112: Plank

Strengthen your ankles and legs. When it comes to strength training, most people think of their arms, chest, and back. However, it's the legs that really count. Your leg muscles are among the largest and most powerful in your body. It's their job to help you counteract gravity. Having strong ankles and legs will provide much more benefit than upper body conditioning. Two simple exercises will do the trick.

First, hold on to something for stability and stand up on your toes. Do this at least ten times every day. Work up to twenty repetitions.

Second, add squats to your daily routine. This is perhaps the most important strength exercise you can do. It will decrease your risk of falls and help your overall fitness. Just be sure to loosen up before trying them and maintain good form. Work up to twenty squats every day.

Fig. 113: Foot exercises and squats

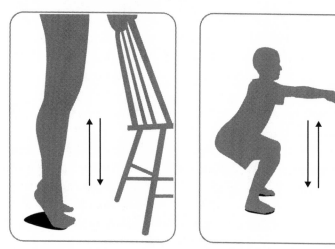

Strengthen your chest, arms, and back. Legs may be more important, but let's not overlook our upper body. You're certainly welcome to use weights and machines if you're comfortable, but they really aren't necessary. Spend a few dollars on a resistance band and do this pull-apart exercise every day (see figure 114). It's another exercise that you can easily do while watching TV.

As your fitness level improves, you might consider working in two more simple but very effective exercises—push-ups and pull-ups. With push-ups, you can start just by leaning against a bed, wall, or chair. Don't try to perform them from the horizontal position until you're sure you're ready. As with all these recommendations, a little bit done consistently is far more effective than a lot done intermittently.

Fig. 114: Chest exercise

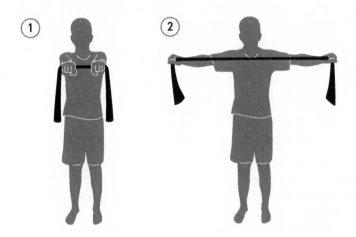

When you're ready for some serious back and shoulder exercises, try to start doing pull-ups. You can easily buy a pull-up bar that you can mount in any doorway. Start with your feet resting on a chair. Work up to twenty or so pull-ups a day this way until you're strong enough to perform one without the aid of the chair. Depending upon your age and fitness level, you can increase the number to whatever you like, but even two sets of four or five pull-ups per day is awesome.

Improve your grip strength. The strength of your grip might seem unimportant, but it is an excellent indicator of health and fitness. Keep a tennis ball in your car or office and squeeze it twenty times using each hand every day. Try to hold the squeeze for at least five seconds. You'll find that your hand strength increases rapidly and helps in many daily activities.

Work these foods into your daily diet. Again, what you eliminate from your diet is less important than what you include. Here are some plant-based substances that you should try to consume every day.

- Ground flaxseed. Flax is a great source of lignans. Try to consume two tablespoons every day. You can put ground flax into a smoothie, pancakes, oatmeal, or baked goods. You can even sprinkle it over a salad.

- Turmeric. Turmeric is a root that contains a number of powerful phytochemicals. Chief among them is curcumin, a potent antioxidant and anti-inflammatory. Grind a teaspoonful each day and add to various types of foods—pasta sauces, salads, casseroles, and so on.

- Berries. Blueberries, strawberries, blackberries, and raspberries are all pharmaceutical miracles. They act not only as antioxidants but also as mild mitochondrial poisons that elicit a protective response. Consume as much as possible. They're great in muffins, pancakes, waffles, breads, smoothies, and many other dishes.

- Leafy greens. Spinach and other leafy greens (lettuce, arugula, chard, etc.) have high levels of B vitamins and folate along with beneficial phytochemicals. They should be a daily part of your diet, in salads, smoothies, or all by themselves.

- Broccoli and other cruciferous vegetables. Cruciferous vegetables contain a phytochemical called sulforaphane that has a number of beneficial effects. We talked about the protein transcription factor NRF2 in the chapter on cell signaling. These vegetables activate NRF2, which in term ramps up a number of protective mechanisms within the cell. If you really want to consume large amounts of sulforaphane, consider eating broccoli sprouts. You can even grow them at home—online explanations of how to do so are readily available.

Get seven to eight hours of high-quality sleep every night.
Sleep probably deserves its own chapter. Matthew Walker's book
Why We Sleep is highly recommended. Suffice to say that sufficient
high-quality sleep is vitally important to good health. Insufficient
sleep is associated with a variety of diseases from diabetes to Alzheimer's. Take sleep as seriously as you do nutrition or exercise if
you want to optimize your long-term health.

Practice meditation or another form of mindfulness. Modern
life is stressful and filled with things that can occupy our attention. It's more difficult than ever to quiet the mind, but doing so is
immensely beneficial. Techniques like meditation, yoga, focused
breathing, and prayer can help reduce the baseline activation of
the hypothalamic-pituitary-adrenal (HPA) axis and the sympathetic nervous system. As with so many of the recommendations
we've discussed, regular mindfulness can increase overall fitness
and build reserves for the inevitable stressors sure to come.

Consider taking the following supplements. It's easy to go over-
board with supplements. Most are a waste of money and have
been demonstrated to have no overall effect. In particular, avoid
products advertised as "antioxidant." Here are some tried and
true supplements to try.

- Aspirin is a simple way to lower baseline inflammation. However, this is one you should talk to your doctor about. Avoid
  aspirin if you have a bleeding disorder.

- Vitamin D is essential for good immune function. Consider
  10,000 IU per day.

- Fish oil is a great way to get important omega-3 fatty acids that
  reduce inflammation and improve heart health.

- Coenzyme Q is a critical cofactor in the electron transport
  chain—especially important if you're taking statin drugs for
  cholesterol control.

- B vitamins play an important role in a variety of metabolic reactions.

- Sirtuin activators—we've seen the importance of SIRT1 and SIRT3—keep DNA properly wound around histones and also participate in DNA repair. Phytochemicals such as resveratrol and pterostilbene can increase sirtuin activity.

- NAD precursors can help boost NAD levels, which tend to drop with age. Because both the sirtuins and another class of DNA repair enzymes (PARPs) require NAD as a cofactor, consider taking NAD precursors such as nicotinamide riboside (NR) or nicotinamide mononucleotide (NMN).

- Magnesium is a metal that helps many proteins in the cell do their jobs. In particular, it is an important component of several complexes in the electron transport chain. It helps prevent the cell from suffering oxidative stress.

Reduce your body's load of iron and other oxidative metals. Metals like iron are double-edged swords. Our cells still use iron, copper, and manganese at the core of enzymes in the electron transport chain—an ancient vestige of our mineral origins. However, these metals usually reside in protein cages where cells can tap into their benefit without incurring damage. When roaming free, iron in particular can combine with oxygen to create dangerous free radicals. Men especially should keep iron intake to a minimum without becoming anemic. Enjoy red meat in moderation because of the iron content.

Engage socially. Humans are social animals. We need interpersonal contact to be psychologically and physically healthy. Make a point of maximizing your human interaction every day. Of course, talk to friends and family as much as possible. However, you can engage others in your day-to-day activities as well. Smile. Ask the checkout clerk at the grocery how they're doing. Volunteer. There are countless opportunities.

Exercise your mind. Mental stimulation is key to building a large cognitive reserve. Studies consistently find that those with more mental stimulation in middle age have a lower risk of dementia as they get older. Learn a new language. Take up a new hobby. Read instead of watching TV—or listen to audiobooks.

Have a purpose. It's amazing how our health is influenced by our overall mental outlook. Key to that outlook is our sense of purpose. Older people who lose a long-time spouse whose care or companionship was their primary purpose often die soon after. It's critical both to have a purpose and to consciously know what it is. This knowledge helps your brain send signals to the rest of the body that say, "I'm still useful. I'm not ready to go yet!"

Laugh and try not to take it all too seriously. No matter how hard we try, we can't hang on to life forever. Paradoxically, obsessing too much about health can actually be unhealthy. What most of us want is not to live forever but rather to be in a position to enjoy whatever time we do have. Long-lived people have an incredible way of laughing at life while enjoying it at the same time.

Know Your Body

The ancient Greek physician Hippocrates said, "If you are not your own doctor, you are a fool." This sentiment is especially true today. The volume of information we can generate for a person's health is enormous. There's simply no way for your physician to have the time or expertise to adequately follow every aspect of your health. We recommend that you treat your physician as a helpful guide rather than turn over to them the keys to your health. Now you can run your own tests and evaluate your own health—with the help of your physician. We recommend that you perform the following tests every year, analyze the results, and ask your physician about anything you don't understand. Companies will perform these tests with no doctor's order required.

Blood Test

These biomarkers can be tested for roughly one hundred dollars.

Metabolic function
- Insulin
- Fasting glucose
- Hemoglobin A1c

Inflammation
- High sensitivity C-reactive protein (hs-CRP)
- Homocysteine
- Sedimentation rate

Lipids
- Cholesterol
- LDL
- HDL
- Triglycerides

Vitamins
- Vitamin D
- Vitamin B12

Liver function
- Albumin
- Alkaline phosphatase
- ALT
- AST
- Bilirubin
- Total protein

Kidney function
- Blood urea nitrogen (BUN)
- Creatinine
- Glomerular filtration rate (GFR)

Electrolytes—CO_2, chloride, potassium, sodium

Red blood cells—platelet count, hemoglobin, hematocrit, RBC count

White blood cells—total count, basophils, eosinophils, lympho-cytes, monocytes, neutrophils

Immune function

Endocrine
* Thyroid stimulating hormone (TSH)
* T3, T4

There are other blood tests you might want to consider running:

* Iron—% saturation, ferritin, iron binding capacity, total iron, transferrin
* Minerals—copper, magnesium, zinc
* Autoimmunity—anti-nuclear antibody (ANA)
* Fasting cortisol
* Fat hormones—leptin and adiponectin
* Sex hormones—testosterone, estrogen, progesterone

Physiological Measurements

These are quantities you can measure yourself with some simple equipment:

* Height
* Weight
* BMI (calculators online; all you need is your height and weight)
* Blood pressure
* Resting heart rate
* Lung volumes
* Body fat percentage
* Grip strength

Genetic Tests

You can get your entire genome sequenced for about five hundred dollars. However, the most important aspect of your genotype that you should know is which versions of the ApoE gene you carry. There are three common variants, called E2, E3, and E4. E4 is associated with a greatly increased chance of developing

Alzheimer's—especially if you have two copies (one from mom, one from dad). If that's the case, you should see an Alzheimer's specialist immediately and adopt a rigorous lifestyle plan that mostly involves following the recommendations in this chapter. The E3 variant is neutral, while the relatively rare E2 version actually reduces your risk of developing dementia. Some versions of the 23andMe genetic test return this result.

Cancer Tests

Consider having a whole-body MRI. This a relatively low-cost way to scan for cancer that doesn't subject you to a lot of radiation. There's some controversy around this approach, however. Some physicians feel that whole-body MRIs uncover too many false positives and lead to investigation of benign problems. But they can also find early cancers that might otherwise have been lethal. In the author's estimation, knowledge is power. The more we can have of it, the better.

A rapidly emerging way to screen for cancer is to look in the blood for circulating tumor cells (CTCs). Cancer cells frequently slough off tumors and enter the blood stream. DNA sequencing techniques are now sensitive enough to detect these tumor cells. Researchers were surprised to find that it's possible to find signs of early cancers in the blood. Soon it will be possible to have an annual blood test to check for the presence of cancer anywhere in the body.

Heart Tests

The cardiac CT uses X-rays to create a three-dimensional view of the heart and major blood vessels. This test can be used to detect blockages in the coronary arteries that supply blood to the heart wall itself. It can also measure the overall amount of atherosclerosis throughout the heart, resulting in the so-called calcium score that reflects the amount of plaque buildup in the heart. This is a test that everyone should consider as they move into their fifties.

Other Tests

Tests are emerging that can easily look for some of the proteins associated with Alzheimer's. This is especially important for those with a family history of dementia or who are known to carry the E4 version of the ApoE gene. One promising test examines the ratio of two different ways of cutting the amyloid precursor protein (APP) by measuring the amount of one amyloid beta fragment forty-two amino acids long and another one forty amino acids long. A low ratio of Aβ42/40 indicates elevated risk for Alzheimer's.

Monitor Your Progress

Consider repeating certain blood tests every six months or so. This will allow you to see how much progress you're making. Keep in mind that these are your high-level goals:

1 **Maintain good insulin sensitivity.** This will generally be reflected by your insulin level. High levels mean that your body is having to produce a lot of insulin to get your muscle and other tissues to even pay attention. An insulin level of five uIU/mL or less means that your insulin sensitivity is very good.

2 **Keep inflammation low.** You should work to bring your hs-CRP level down below 1.0 mg/L. Other markers of inflammation such as homocysteine and sed rate should be low as well.

3 **Develop healthy lipids.** Your focus should be on building your HDL ("good cholesterol") levels as high as possible while keeping LDL ("bad cholesterol") under 100 mg/dL.

4 **Optimize cardiovascular health.** You can get a good idea of the status of your cardiovascular system from two values—resting heart rate and blood pressure. The absolute value of resting heart rate is less important than how it changes over time. You'd love to see yours go down as you get fitter. As for

blood pressure, you should aim for a systolic pressure (top number) under 120 mm Hg and a diastolic pressure (bottom number) under 80 mm Hg.

The Most Powerful Force in the Universe is Compound Interest

Don't fall into the trap that ensnares so many people at the beginning of the year. Everybody wants to turn all the knobs at once to get fit and healthy. Unfortunately, that approach doesn't work for most of us. And we see that reflected in the statistics every year, showing that most people have given up on their "big bang" programs before the end of January.

Your best chance of success is through small changes adopted so consistently that they become lifelong habits. For example, instead of resolving to jog an hour a day, make sure that you spend fifteen minutes every evening stepping up and down from a small step—while watching TV if you like. By all means, try to go vegetarian if you want, but start off by trying to get five servings of healthy fruits, vegetables, seeds, nuts, and oils into your body every day. And if you desire to develop muscle mass by pumping iron in the gym three times per week, go right ahead. But start off by trying to consistently perform the toe stands, planks, and push-ups recommended in this chapter.

Instead of sprinting into the new year trying to instantly redo your entire life, pick some small changes that you *know* you can maintain. In general, avoid short-term changes like the plague. Special diets, quick-fix exercise programs, and other short-sighted behavioral changes don't work for most people. Make small, incremental changes and turn them into lifelong habits. It's the small things you do consistently that make the biggest difference in the long run.

References and Further Reading

D Aune et al (2017). Fruit and vegetable intake and the risk of cardiovascular disease, total cancer and all-cause mortality—a systematic review and dose-response meta-analysis of prospective studies. *International Journal of Epidemiology*, 46(3): 1029–56.

JD Creswell and EK Lindsay (2014). How does mindfulness training affect health? A mindfulness stress buffering account. *Current Directions in Psychological Science*, 23(6): 401–7.

W Kemmler and S von Stengel (2011). Exercise frequency, health risk factors, and diseases of the elderly. *Archives of Physical Medicine and Rehabilitation*, 94(11): 2046–53.

DP Leong et al (2015). Prognostic value of grip strength: findings from the Prospective Urban Rural Epidemiology (PURE) study. *Lancet*, 386(9990): 266–73.

AM Minihane, S Vinoy, WR Russell, A Baka (2015). Low-grade inflammation, diet composition and health: current research evidence and its translation. *British Journal of Nutrition*, 114(7): 999–1012.

JE Turner, VA Lira, PC Brum (2017). New insights into the benefits of physical activity and exercise for aging and chronic disease. *Oxidative Medicine and Cellular Longevity*, Article ID 2503767.

17

What is Life?

"Thus the device by which an organism maintains itself
stationary at a fairly high level of orderliness really consists of
continually sucking orderliness from the environment."

ERWIN SCHRÖDINGER

WHAT EXACTLY is life? Humans have been asking themselves this question since the dawn of time. On one level,
the question is trivial. Living things are different from
inanimate objects in fundamental ways—they move, change,
and interact with their environment. But thinkers great and varied, from Aristotle to Schrödinger, have tried to figure out exactly
what makes one thing alive and another not.

The French mathematician and scientist René Descartes had a
large hand in synthesizing the modern view of life. To Descartes,
living things were automata: machines made of material components and completely separate from the mind, which existed only
in humans. This mechanistic view of the body and the mind-body
duality greatly influenced Western science, right up to the present day. It led most scientists to think of the body as completely
material and the mind as something else altogether, the "ghost in

the machine." This outlook fit well with Newton's discovery of the laws of motion, which seemed to explain the physics of the universe.

As we move toward the middle of the twenty-first century, however, a quiet revolution is brewing. Physics has been shaken to its core by the discovery of quantum mechanics and its attendant "weirdness." Instead of consciousness as something to be held to the side, the notion of an intelligent observer is assuming a central, although still uncomfortable, position. The noted English physicist James Jeans summed up the situation with the memorable line: "The universe begins to look more like a great thought than like a great machine." Other modern physicists see our universe as a projection from a timeless two-dimensional plane, hearkening back to Plato's allegory of people chained in a cave who watch shadows projected by a fire behind them and think of the shadows as reality.

After spending centuries throwing off the intellectual shackles imposed by organized religion, modern Western scientists are loath to allow anything that can't be observed and measured into the canon. Yet the unshakeable role of consciousness in the quantum world has forced many to face up to the undeniable implications. Biology is finally having its own "observer effect" moment. What if life isn't a simple accident involving the random assembly of different molecules that finally began to replicate? As in physics, what if there are deeper forces at play?

This certainly doesn't have to mean dialing back the clock a few centuries and returning to a human-like deity who meticulously designs every aspect of creation. Anyone who has ever studied DNA from living organisms can see that it's as far from intelligent design as could possibly be imagined. For a professional software developer, the "code" in DNA looks like an absolute mess—"spaghetti code" in the vernacular. What's miraculous is that it works at all. We see clear evidence of numerous copy-paste events with the new code evolving slowly away from the original. There are enough side effects and crosstalk to make an engineer shudder. As odd as it may seem, the only sign of perfection in life is in the process of evolution.

A new view of life and the universe itself is emerging based on information theory. Physicists have always viewed information as arising from changes in physical quantities, such as the changing vocal sounds produced by vibrating air molecules in speech. However, some theorists are flipping this conception on its head, basing reality on information rather than matter.

This information-first view has the potential to unify physics and biology. Increasingly, we're starting to view life as an information-processing system. It would be no exaggeration to think of life as using energy to extract information from the environment. Life uses this information to find nutrition, adapt to current conditions, and even to anticipate future changes.

One of the key features of life is that it relies on templates. Of course, we see this most directly in DNA. Each strand of a DNA molecule is a template for its sister strand. The sections of DNA we call genes act as templates to make messenger RNA that carries protein recipes to the ribosome for fabrication of the specified protein. In thinking about the origin of life, we can well ask, what was the original template that gave rise to the first genetic material?

As we learned in the first few chapters of this book, life developed from minerals. Like the complex repeating molecules in living things (DNA, proteins, etc.), minerals form sequences. In the case of the diamond, the sequence is a monotonous series of carbon atoms broken only by random imperfections. Other minerals have a wider variety of composition, leading to interesting patterns when the minerals form. Looking at figure 115, it's not hard to see how a chain of DNA or RNA molecules (collectively referred to as nucleic acids) could develop on top of a mineral surface with the chemical composition of that chain somehow reflecting the variation in the mineral surface.

Could it be that the sequence of one nucleic acid chain (think DNA or RNA) formed in this way just happened to be able to copy itself? The exact structure of a given crystal reflects both the laws of physics and the conditions that happened to exist when that crystal was formed. So, in a real sense, any nucleic acid reflecting the structure of that crystal represents information drawn from it. Did life read a hidden secret buried in the rocks that jump-started the whole process? We've already seen how rocks helped catalyze the essential reactions that provided energy to early life. Perhaps they also provided the initial genetic blueprint to set the entire wheel in motion.

Fig. 115: Mineral surface

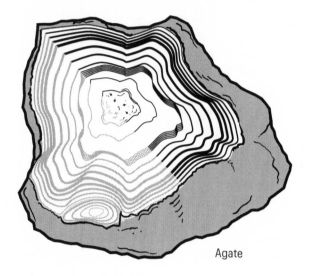

Agate

The Future

As our understanding of life grows at all levels—molecule, cell, organism—we're starting to identify steps we can take to reduce the rate of aging and sometimes even reverse its effects. Here are some of the exciting trends to keep an eye on.

Senolytics

Our cells have a built-in quality control mechanism that will keep them from dividing if a problem is detected. A variety of conditions will trigger this mechanism, including DNA damage, short telomeres, and an overly rapid cell cycle. This system is thought to have evolved to protect against the development of cancer. When shutdown occurs, the cell continues to live but is no longer able to divide. Such zombie-like cells are called *senescent*.

In youth, we have very few of these cells, and our immune system recognizes and destroys them quickly. As we age, however, more cells trigger the senescence shutdown, and our slowing immune system is unable to catch them all. The number of senescent cells rises exponentially, creating inflammation and weakening tissues throughout the body. A few years ago, a team at the University of Wisconsin demonstrated that killing off senescent cells had an impressively rejuvenating effect on laboratory animals.

Fig. 116: Mice

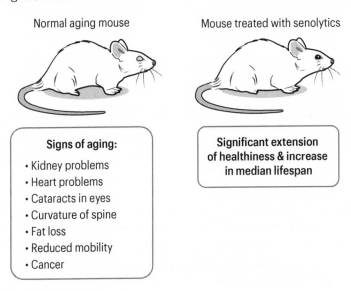

Normal aging mouse

Mouse treated with senolytics

Signs of aging:
- Kidney problems
- Heart problems
- Cataracts in eyes
- Curvature of spine
- Fat loss
- Reduced mobility
- Cancer

Significant extension of healthiness & increase in median lifespan

Source: Baker (2011)

Companies around the world are now racing to develop medicines that act as *senolytics*—they lyse (kill) senescent cells, allowing stem cells in the corresponding tissues to replace them. At this point, it's unclear how dramatic the effect will be in humans, but senolytic drugs seem destined to be part of the anti-aging arsenal.

Stem Cells

Every tissue in our body has a small supply of stem cells—special cells that divide to replace other cells lost due to death or injury. Some tissues have more stem cells than others. For example, stem cells in the skin and intestines continue to divide and maintain their tissues for over one hundred years. Other tissues such as the heart and brain develop early and don't have the same sort of constant turnover as the skin and intestinal tract. As a result, they have very few stem cells and don't recover well from injury.

In 2006, Japanese researcher Shinya Yamanaka announced that he had been able to turn fully differentiated skin cells back into stem cells just like those in the embryo. These stem cells could form any tissue in the body, including brain and heart. And the process of "reprogramming" differentiated cells back into stem cells was shockingly simple, requiring just four protein transcription factors.

Stem cells are slowly starting to make their way into clinical practice and seem destined to become commonplace. At least in theory, they will allow us to replace lost cells anywhere in the body. This could mean re-establishing cartilage in worn-out joints, growing new heart muscle in heart attack patients, and perhaps even replacing neurons lost due to Alzheimer's or Parkinson's. Interestingly, one team has found that much of aging can be traced to loss of function in the hypothalamus—a tiny part of the brain that regulates temperature, daily cycles, and many other aspects of physiology. We may be able to turn back the clock by injecting stem cells into the brains of older people.

DNA Editing

As we age, our DNA accumulates more and more mutations. Recent years have seen the discovery of remarkable ways to accurately edit the DNA within cells. This holds the prospect of being able to correct those errors using DNA editing tools such as CRISPR/Cas9. One particularly exciting possibility is to edit errors in hematopoietic stem cells (HSCs). These are stem cells that live in the bone marrow and give rise to all the cellular components of the blood—red blood cells, white blood cells, platelets, etc. Doctors can already use radiation to kill the bone marrow and blood cells of patients with leukemia and other disorders, then inject HSCs to regenerate the whole system. Someday soon, we may be able to remove HSCs from a patient, use CRISPR to fix any DNA mutations, then re-inject them back into the patient. The net effect would be to rejuvenate a hugely important subset of the body's cells.

Epigenetics

Many of the changes in the proteins made by cells as we age are not due to mutations in the DNA but changes in how the DNA recipes are read. Eukaryotic cells have evolved elaborate mechanisms to control which genes are active or inactive within a given type of cell. This can involve attaching molecules to the histone protein spools that DNA is wrapped around to control the tightness of that wrapping. The DNA can also be marked directly by attaching methyl ($-CH_3$) groups to certain DNA letters, which has the effect of inactivating any nearby gene.

Epigenetics—the study of heritable changes in gene function that do not involve changes in DNA sequence—is starting to play multiple roles in research on aging. For one, we can analyze the epigenetic marks and use them to calculate a person's "biological" age—that is, functional age rather than chronological age. As we get older, the wrapping of DNA around histone spools becomes less and less exact. In some parts of the genome, the DNA becomes less tightly wrapped, allowing genes in the area to be used to make proteins when they shouldn't be making them. There's great potential for epigenetic interventions that can restore the proper packing of DNA and other therapies that might help us live longer and more healthily.

Conclusion

It's incredible that aging wasn't even a recognized topic of research until recently. People, including scientists, just assumed that aging was an inevitable process over which we had no control. They viewed growing old as the result of gradual wear and tear ultimately leading to a breakdown. Our understanding of how cells—and consequently our own bodies—work has exploded in the last three decades. The precipitous drop in the cost of genetic sequencing has allowed us to inspect the sacred DNA-based recipe books in our cells and compare them to those in other creatures from bacteria to baboons.

The results have been shocking. All our books contain largely the same recipes—the same kinds of proteins and how to make them. Just as ancient manuscripts copied by hand contain a few errors, so, too, do the DNA recipe books as different creatures diverge. However, we can compare the various versions of a recipe and clearly see that they all started from the same original. And it's not just individual recipes that have been passed down through the eons. Entire sets of genes that implement complex behaviors referred to as "programs" have been passed down too.

We'll be studying our DNA and that of our evolutionary cousins for decades to come. However, we know enough already to draw certain conclusions. Cells must constantly evaluate their internal operations as well as the environment in which they live. They must determine when conditions are good for growth and reproduction. Just as importantly, they must recognize impending problems and react before it's too late. And since early trouble is often indicative of even worse to come, cells must work to get ahead of potential problems. For that reason, they tend to overreact to mild stress in anticipation of more severe conditions just around the corner. It is becoming clear that activating these mild stress response programs increases our fitness and our overall ability to handle whatever the universe throws at us.

The thrilling realization is that humans were made to move, change, and adapt. We are the species that walked out of Africa and spread to every corner of the globe. Our ancestors survived blazing deserts, hazardous mountains, blinding snowstorms, and every condition imaginable. They endured long periods without food and great journeys full of uncertainty. Over the millennia, our cells, tissues, and bodies have learned to interpret these discomforts as *purpose*. The only reason to suffer through such pain and privation is to accomplish something important—escaping a dangerous enemy, finding a new hunting ground, establishing a new home for our children, and so on. This may sound brutal, but when we sit around in warmth and comfort, it's time to die. To nature, a lack of challenges is a powerful signal that we are no longer needed.

The key takeaway is this: Make yourself uncomfortable. Hike in the snow. Run in the rain. Laugh at the comedy of life and cry during times of sorrow. Climb a mountain, bathe in a hot spring, shiver in an icy river. Explore your world to its fullest and embrace change and uncertainty. We are the kin of mighty explorers who sacrificed the comforts of hearth and home for the thrill of discovery. We settle into our rocking chairs at our peril. You—and the rest of us—were meant for greater things.

References and Further Reading

AG Cairns-Smith, AJ Hall, and MJ Russell (1992). Mineral theories of the origin of life and an iron sulfide example. *Origins of Life and Evolution of the Biosphere*, 22: 161–80.

P Genovese et al (2014). Targeted genome editing in human repopulating haematopoietic stem cells. *Nature*, 510: 235–40.

S Pal and JK Tyler (2016). Epigenetics and aging. *Science Advances*, 2(7): e1600584.

LF Seoanne and RV Solé (2018). Information theory, predictability, and the emergence of complex life. *Royal Society Open Science*, 5(2).

SI Walker et al (2017). *From Matter to Life*. Cambridge University Press.

M Xu, T Pirtskhalava, and JL Kirland (2018). Senolytics improve physical function and increase lifespan in old age. *Nature Medicine*, 24: 1246–56.

Index

About the Author

AFTER EARNING an undergraduate degree in physics, Don Brown enrolled in a combined degree program at the Indiana University School of Medicine. Under that program, he was awarded a master's degree in computer science in 1983 and an MD in 1985. Three decades later, he went back to school and earned a master's degree in biotechnology from Johns Hopkins in 2017.

Don is one of the most successful serial entrepreneurs in the Midwest. While finishing medical school, he started his first company, which was acquired by Electronic Data Systems in 1986. He then founded Software Artistry, which became the first software company ever to go public in the state of Indiana, and which was later acquired by IBM for $200 million. Don then founded Interactive Intelligence, which went public in 1999 and was acquired by Genesys Telecommunications Laboratories in 2016 for $1.4 billion.

Shortly after the Genesys acquisition, Don founded LifeOmic to bridge his passion for the life sciences with his experience in cloud technologies and artificial intelligence. LifeOmic built a precision health software platform to help researchers and clinicians leverage the power of big data to identify the mutations driving cancer and provide personalized treatment. LifeOmic has also developed two award-winning mobile wellness applications used by millions of people around the world along with a next-generation corporate wellness solution.

Don is an active technology investor and philanthropist. In 2016, he donated $30 million for the establishment of the Brown Immunotherapy Center at the Indiana University School of Medicine.

Don is an avid outdoorsman who loves hiking, rock climbing, and skiing with his eight children in and around Park City, Utah. He was named Sagamore of the Wabash by Indiana governor Mitch Daniels in 2012.

Made in the USA
Monee, IL
01 February 2022

90443189R00182